NEW WORKS IN ACCOUNTING HISTORY

edited by

RICHARD P. BRIEF
LEONARD N. STERN SCHOOL OF BUSINESS
NEW YORK UNIVERSITY

TRADE ASSOCIATIONS AND UNIFORM COSTING IN THE BRITISH PRINTING INDUSTRY, 1900–1963

STEPHEN P. WALKER
FALCONER MITCHELL

Routledge
Taylor & Francis Group
New York   London

First published 1997 by Garland Publishing, Inc.

This edition published 2013 by Routledge

Routledge
Taylor & Francis Group
711 Third Avenue
New York, NY 10017

Routledge
Taylor & Francis Group
2 Park Square, Milton Park
Abingdon, Oxon OX14 4RN

Routledge is an imprint of the Taylor & Francis Group, an informa business

Library of Congress Cataloging-in-Publication Data

Walker, Stephen P.
    Trade associations and uniform costing in the British printing industry, 1900–1963 / Stephen P. Walker, Falconer Mitchell.
        p.   cm. — (New works in accounting history)
    Includes bibliographical references and index.
    ISBN 0-8153-3024-3 (alk. paper)
    1. Printing industry—England—Accounting. 2. Printing—England—Costs. 3. Printing—England—Societies, etc. 4. Printing industry—England—Cost control. I. Mitchell, Falconer. II. Title. III. Series.
HF5686.P8W35 1997
651'.867—dc21   97-23319

CONTENTS

ACKNOWLEDGMENTS

The authors are grateful to the British Printing Industries Federation and The Trustees of the National Library of Scotland for permission to quote from and reproduce material contained in this book. Thanks are also due to Douglas McRae, David Padbury and Nigel Roche for their helpful guidance in locating source materials and to Professor Richard P. Brief for suggestions which have greatly improved the content and appearance of the volume. The authors gratefully acknowledge the financial support of the Chartered Institute of Management Accountants (CIMA).

INTRODUCTION

This book documents a highly significant development in the history of costing practice in the UK - the uniform costing system designed for the members of the British Federation of Master Printers (BFMP) during the early twentieth century. Shortly after the launch of the printers' system, industry based uniform costing schemes became popular in the UK. From 1913 to 1939, 26 such systems appeared, covering key industrial sectors including rubber, paper, electrical, iron and steel, confectionery, clothing, plastics, cotton and paint (Solomons, 1950a, pp. 241-42). Thus uniform costing technology was potentially a prime influence on early twentieth century costing practice in several industries. The uniform costing system devised for the British printing industry was the earliest and proved to be the most enduring (even today costing manuals are published by the British Printing Industries Federation). Although the printers' uniform costing system was officially launched at 'The First British Cost Congress' in 1913 this event was preceded by a decade of costing activity by the BFMP which resulted in the publication of several pamphlets on the subject (Mitchell and Walker, 1997). This chronology places the uniform early costing work of the printers among the first practical publications on costing to receive widespread circulation in Britain (for example the BFMP's 1904 publication on costing, *Profit for Printers*, went into second edition in 1907 following the exhaustion of the 7,500 copies of the first edition).

The case of uniform costing in the British printing industry assumes significance in the history of industrial accounting for a number of reasons. The episode contributes to the expanding literature in accounting history which addresses issues concerning the motivations for the imposition of early costing systems (see Fleischman, 1996; Fleischman et al., 1996, pp. 66-69). The instance presented here emphasises the need to encompass craft-dominated industries in the discourses together with the impact of the interposition of a trade association on the emergence and application of a costing technique.

The printers' case illustrates how adverse economic circumstances and a desire by employers to advance their collective socio-economic and political status appears to have driven best practice. It shows how market contingencies dictated the need for full as opposed to direct costing, and for the inclusion of depreciation and interest on capital in the system of cost determination (the contemporary treatment of both was, in general, somewhat arbitrary). These practical choices were made to ensure that the product costs which were determined under the uniform costing system were as comprehensive and as high as possible. The intended impact of this policy on supplies was to maintain industry prices at an elevated level and thereby increase the profits of employer printers.

The BFMP case is also important due to the influence which the promoters of uniform costing had in other sectors. The minute books of the Costing Committee of the BFMP reveal that the printers' scheme was emulated during the 1920s by trade associations representing master bookbinders, paper box manufacturers and paper bag makers. In the 1930s associations of india rubber manufacturers, wallpaper manufacturers, wholesale clothing manufacturers and tin box manufacturers were similarly interested. The costing publications of the BFMP were widely circulated and commented upon outside of the printing industry. The leading advocates of uniform costing in printing promoted the existence of the BFMP's scheme to interested parties such as professional accountants and students (see *The Accountant*, 28.2.1914, pp. 314-17; 10.12.1927, pp. 783-84).

Certain of the leaders of the uniform costing movement in Britain were also involved in the professionalisation of cost accounting (Mitchell and Walker, 1995, p. 37). W. Howard Hazell (Chairman of the Costing Committee from 1918-1929 and a President of the BFMP), A.E. Goodwin (Secretary of the BFMP Costing Committee 1913-17 and a Secretary of BFMP) and A. Williamson (Secretary of the BFMP Costing Committee 1917-33) were members of the inaugural council of the Institute of Cost and Works Accountants (ICWA) which was formed in 1919. Hazell was the first Vice-president and second president (1925-29) of ICWA and A. Williamson was the first Chairman of the ICWA Membership and Branch Committee and also served on the ICWA's Executive and Finance Committee.

THE TRADE ASSOCIATION: AN OVERVIEW OF ITS DEVELOPMENT AND RATIONALE

Trade associations have a long history in the UK. Their antecedents may be traced to the system of trade guilds which existed in the Middle Ages. Trade associations were reasonably common by the mid-nineteenth century (Farnham and Pimlott, 1990) but at that time were organised on a local or regional basis and their operation was usually spasmodic and designed to cope with localised labour disputes. Early trade associations in printing were typical of this. The London printing trade, for example, had a history of spasmodic attempts to negotiate and regulate rates of pay based on a scale devised in 1810 and amended by a Conference of Master Printers and Compositors in 1847 (Clegg, 1976, p. 123).

Trade associations which were organised on a national or federal basis began to feature more prominently during the late nineteenth and early twentieth centuries. This development is usually explained as a component of the 'employers' offensive' or 'backlash' of the 1890s which comprised a reaction to the combination of unskilled labour and the 'new unionism' movement. The Great London Dock Strike of 1889 had been broken by the Shipowners Federation. In response to this defeat trade unions improved their organisation, developed a more extensive membership and became more politically active and more militant (Palmer, 1983; Clegg, 1976). The threat posed to employers by the combination of labour was countered by the formation of national organisations, engaging the unions in disputes, lobbying government (an Employers Parliamentary Council was formed in 1897) and by instituting legal challenges to trade union rights. The latter culminated in the Taff Vale judgement of 1901 (Clegg et al, 1964, pp. 126-151). In 1895 there were about 335 employers' associations in Britain. By 1914 there were almost 1,500 (Gospel and Littler, 1983).

While the threat of organised labour was one important factor which encouraged the emergence of national trade associations, there were other significant reasons for the advance of employer organisations. A national trade association could also provide valuable specialist advisory services for its members. These included the interpretation of relevant legislation, training and educational provision for management, information on health and safety matters, the provision of legal advice, job evaluation and work study, and the distribution of comparative statistics on various aspects of performance. In addition, the association could function as a pressure group and assume a representative role in order to protect the interests of its members and secure political and economic advantages. As an organisation representing those responsible for the supply of a product or service, the trade association was in a strong position to influence and co-ordinate market behaviour to the benefit of its members. Indeed by 1919 this role had been explicitly recognised by government:

> We find that there is at the present time in every important branch of industry in the United Kingdom an increasing tendency to the formation of Trade Associations and Combinations having for their purpose the restriction of competition and control of prices. Many of these organisations which have been brought to our notice have been created in the last five years, and by far the greater part of them appear to have come into existence since the end of the century (Ministry of Reconstruction, *Report of the Committee on Trusts*, 1919, Cd. 9236, p. 2, quoted in O'Brien and Swann, 1968, p. 17).

THE MARKET MOTIVE

From the perspective of the individual businesses in an industry, the economic advancement of member firms is one of the key attractions of trade associations. While enhancing employers' strength in trade union negotiations provides one way in which economic benefit can be obtained, this can be complemented by the control (see following section) which a trade association can bring to bear on the product or service market in which its members operate. As a supra-organisational entity the trade association occupies a unique position in attempts to influence market supply and thus the profitability of individual firms. The potential for financial gain from this type of supply control is illustrated in Figure 1.

If the trade association ensures price is maintained at P_1, despite a slump in industry demand meriting a fall to P_0 (i.e. where total industry supply and demand intersect), then clearly, the maintenance of the higher price will reduce the quantity of industry demand and this will be felt by individual firms as a reduction of quantity sold to Q_2. Despite this lower volume of sales, the individual producer's profit contribution will be enhanced by the difference in profits earned at Q_0 (P_0CX) and Q_2 (P_1ADX) (that is, the excess of rectangle P_1ABP^0 over triangle BCD). Thus the trade association's control of price has benefitted the individual firm financially and has had the additional attraction of increasing price stability. However, the association's actions have also created a situation where the individual firm has a motivation to break rank as by so doing its own profits can be substantially increased. In the above case, with a price of P_1, the individual producer would be tempted to sell up to Q_1 and so enhance profits.

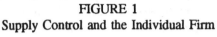

FIGURE 1
Supply Control and the Individual Firm

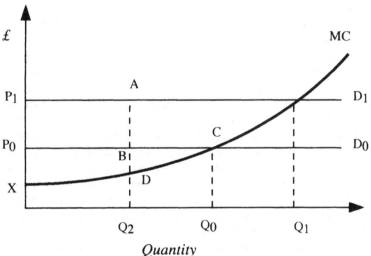

Quantity

Of course, if all firms acted in this way then none would gain as the market would be flooded and (unless price fell) there would be a proliferation of unsold output. Moreover, the individual producer might find it beneficial to reduce price (to any level between P_1 and P_0) when the resultant extra sales volume more than compensated for the revenue lost from the lower price. The success of supply control therefore depends upon the willingness of members to restrict output and, on an individual basis, to forego the temptation of altering price and output to generate higher short-run profits. This constitutes the dilemma and potential instability of collusive pricing, whether tacit or overt. Profits can be collectively enhanced but the very act of enhancement creates opportunities for, and pressures on, individual firms to pursue particularist strategies at the expense of other suppliers.

MECHANISMS FOR SUPPLY CONTROL

The financial benefits of a trade association's influence on supply considerations depend upon the maintenance of a price level which is different to that warranted by the free operation of demand and supply. The trade association can provide a number of mechanisms to facilitate price control. First, it can provide visibility to the pricing policies and practices adopterd by its members. This is termed "open pricing" and takes the form of the full disclosure of the terms of every sales transactions made by individual members. The trade association subsequently organises the rapid dissemination of price data to other members (Scherer, 1981, p. 81). Those members who engage in extensive price cutting are thereby identified and can be subjected to peer pressure to desist from such behaviour or the threat of reprisals from their competitors. Open pricing can be effectively complemented by a second approach to price control - the periodic publication and circulation of standard price lists for the main products of the industry. Although these may not be compulsorily imposed on members they do provide guidance on the pricing expectations of the trade association and if supplemented with 'open pricing' policies, can clearly make visible where price cutting is occurring. The mere presence of these mechanisms may instill commitment among trade association members to follow the 'official' guidance on pricing.

A third approach to price control (which is less explicit than those outlined above) is based on issuing cost of supply information to member firms. This can provide the basis for maintaining price levels either by the overt or tacit consent of member firms to, for example, use cost plus pricing. There are two possibilities here. Average or representative costs for the industry may be computed and circulated to members. Alternatively, guidance may be given to members on how they might discover their own costs. Adherence to full costing principles not only encourages price maintenance but is one aspect of standardisation which facilitates oligopolistic co-operation by increasing the predictability of competitor action and by guiding price setting. For example, awareness of full cost can help to establish resistance points below which price should not fall. These mechanisms will work best in an environment where firms are in regular contact with each other and in this respect the trade association can assist through facilitating meetings and issuing propaganda on the benefits of co-operation between members. As Moore (1970, p. 26) has observed: "An active trade association can often stimulate the recognition of mutual dependence sufficiently for a convention not to cut prices to be observed by more than fifty firms".

HURDLES TO PRICE COLLUSION

While the maintenance of price above the competitive level is one of the greatest attractions of a trade association for member firms, this is not a task which can be easily accomplished. The potential for the individual member to break ranks and increase profits is too strong a motivation for price compliance to occur readily. Furthermore, various aspects of the nature of the market in which the members of a trade association operate can make it more or less likely for a centrally contrived pricing policy to work:

(a)   The more competitive the market the less chance there will be that the artificial price level will be held. Effective collusion is more likely where a small number of large firms comprise an oligopolistic structure. Where the competitive fringe is large and new entrants are attracted in, the trade association will have more difficulty in 'managing' price.

(b)   The more heterogeneous the product or service, the more difficult it will be to establish a general price which is applicable to all suppliers. Opportunities for price shading on an 'under the counter' basis are more common due to the variability of individual sales transactions. Product differentiation provides a justification for price differentiation and this reduces the relevance of central price standards for many individual suppliers.

(c)   The more depressed the economic situation is, the more difficult it is to avoid price cutting. In these conditions the pressure on the individual business to reduce price in order to win business is more intense and adherence to the trade association's pricing guidance runs counter to the pricing policies which may be required for economic survival.

(d)   The higher the fixed cost component of the total costs the more likely is price cutting behaviour. Relatively high fixed costs imply high unit contributions (sales less variable costs) for individual products. The scope for price cuts (at least in the short term) therefore appears to be greater in these circumstances.

Thus the success (particularly in the longer term) of an attempt by a trade association to maintain selling prices for its members at a level which is higher than that which aggregate market conditions would freely determine is likely to be variable. Achievement of this aim will be at least partly dependent upon the varying degrees to which factors such as (a) to (d) above pertain to the sector in which the employers' organisation operates.

THE BRITISH FEDERATION OF MASTER PRINTERS (BFMP)

While the events leading to the formation of the BFMP in 1901 "are wrapped in a degree of obscurity due to the absence of archive material" (Howe, 1950, p. 1), there is considerable evidence to suggest that many of the factors which influenced the general development of national trade associations during the late nineteenth and early twentieth centuries were pertinent to master printers.

The axioms upon which the Federation were based alluded to the advisory support role which a national body could perform and the limited assistance which had previously been provided by the regional associations of printers (see p. 7). The emphasis on the provision of information to members of the trade association centred on three specific areas. First, recent legislation on employment and health and safety at work required summarisation and interpretation in a way which identified their specific operational implications for printers. Second, the industry was undergoing a process of profound and rapid technological innovation (Child, 1967; Musson, 1954; Alford, 1965). From 1889 the introduction of the linotype machine to Britain permitted the replacement of labour-intensive hand composition by mechanical, keyboard operated typesetting. From 1880 rotary press developments and the impact of high speed presses from the United States (after 1900) provided the potential for substantial improvements in quality and productivity. The monotype machine resulted in similar advances in large scale book printing. All this was accompanied by the replacement of steam power by gas and subsequently by electricity.

The new national trade association for master printers thus had extensive opportunities for the dissemination of information to its members on the nature, performance and cost of new technological developments.

The threat posed by organisational advances in printing trade unionism was another important factor which encouraged the emergence of the BFMP. However in contrast to many of the national employers' associations which were created at the turn of the twentieth century, the BFMP did not have overtly aggressive intentions towards organised labour. Their stance, reflecting a history of comparative harmony in industrial relations (Musson, 1954, p. 152), was primarily defensive. On his election as the first President of the BFMP Walter Hazell stated that: "he believed in Trade Unions, but, if they obtained too much power, they were likely to abuse it, and it became essential that there should be Associations of employers to meet the employees on equal terms" (London Members' Circular, May 1901, p. 50). This view was reinforced by the Federation's first Secretary (H. Vane Stow) who saw the creation of a power equilibrium as providing a foundation for sound industrial relations:

> . . . the old proverb "Union is Strength" applies with
> great force to labour questions ... the existence of a
> powerful organisation on the side of the employers
> will tend to that calm consideration of any question
> at issue, which is so necessary to maintain harmony
> between employers and employed of any trade
> (London Members' Circular, June 1901, p. 63).

The competitive nature of the printing industry provided a context which was also conducive to the operation of a system for price maintenance. A national trade association provided the mechanism by which collusive designs could be actualised. At the start of the twentieth century the printing industry benefitted from a number of factors which resulted in a sustained increase in the demand for printed products. During the period of mature industrialisation, public education and mass electoral politics, real incomes were increasing, literacy was improving, consumerism was advancing and industry and commerce were expanding. As a result of these factors the output of newspapers, magazines, cheap editions of books, stationery and other printed products increased substantially. For example in 1883, 3,273 separate newspapers, magazines and journals were produced in Britain. By 1913 the number of titles had increased to 5,555 (Musson, 1954, p. 90). The number of printing firms increased from about 1,200 in 1850 to around 7,000 in 1914 (Alford, 1965, p. 10) and the number employed in the industry increased from 26,000 in 1851 to 152,000 in 1911 (Musson, 1954, p. 90).

A number of sectoral features, however, made it difficult for printing firms to commercially exploit these generally buoyant economic conditions. The industry was structured and operated in a way which fostered competition. While 10% of industrial output was produced by the largest firms (over 200 employees), the bulk of sales (80%) were attributable to medium sized firms (50-200 employees). The remaining businesses comprised small family firms which comprised around 86% of the firms in the industry but accounted for only 10% of the aggregate output of printed products. The fierce competition which emerged between the large and medium printing firms (Alford, 1965, pp. 11-12) was compounded by increasing competition from firms which operated from Continental Europe. Further, as the growth in the number of printing firms during the latter half of the nineteenth century indicates, there were few barriers to the entry of new small scale enterprises in the industry. These smaller firms, through a combination of working long hours, paying low wages (Musson, 1954, pp. 93-94) and operating on the basis of minimal capital investment (Alford, 1965, p. 7), could compete on price for jobbing work. Indeed, as late as 1914 it was the case that "most jobbing work and perhaps most books and better weekly periodicals and magazines were still set by hand" (Musson, 1950, p. 102). The intensity of competition in the printing industry was reflected in falling net profit margins among the larger printers from 25%-30% in 1870 to 10% by 1914 (Alford, 1965, p. 6). This was despite the macro growth in the demand for printing services.

The intensity of the competition in the printing industry during the late nineteenth and early twentieth centuries was also encouraged by the nature of tendering for contracts and one-off jobs. The widespread adoption of blind tendering encouraged printers to enter low quotations to improve their chances of securing custom, and the secrecy of the process avoided any repercussions for the winners from their peers. Customer power derived from the system of tendering was frequently cited by master printers as a justification for price maintenance strategies.

Compounding the problem of customer power in the printing industry was the lamentable state of the cost information which was generated by the majority of price setters. Common deficiencies in printing firms comprised the exclusion of non-production overhead costs from unit cost calculations, the use of general as opposed to departmental overhead absorption bases, the absence of depreciation and the frequent use of arbitrary uplifts to account for indirect costs (Mitchell and Walker, 1996). In 1934-35 William Sessions, President of the BFMP, reflected that:

> . . . in the old days the cost of wages was put down plus the cost of material. The total was then doubled and this was the price of printing. There seemed to be such a large amount of apparent profit in this doubling, so it was easy to cut into! (Howe, 1950, p. 34).

Thus at the time when the BFMP was formed in 1901 fierce price competition was endemic within the British printing industry. The peculiarities of the market constituted a major impediment to the prosperity of employer printers (particularly the owners and managers of the larger firms who dominated the BFMP (*The Caxton Magazine*, June 1923, p. 306)). This state of affairs was made all the more galling by the apparent buoyancy in the overall volume of demand, the exploitation of which was restricted by the structure and conditions in the market. Against this background the BFMP emulated the strategy of American master printers (Powell, 1926) and initiated a programme to develop and propagate a system of uniform costing (Walker and Mitchell, 1996). The aim was to ensure that members were comparably and reliably informed about the full cost of supply when making pricing decisions. This knowledge was intended to introduce more consistency into pricing and would also assist the maintenance of a higher general price level. It was envisaged that when printers were armed with accurate and complete cost data they would be less likely to cut prices to levels which were below the full cost of production.

While the emphasis on price control by the BFMP was a typical feature of trade association activity, the printing sector was not one where the prevailing economic conditions could facilitate the overt operation of this policy. Competition was intense, barriers to entry were minimal, fixed costs were high in the larger capitalised firms and economic downturns were regular and intense during the interwar period. All of these factors created difficulty for the trade association in its attempts to gain the universal acceptance of uniform costing among its members and led to intense efforts over several decades to maintain costing as a practical proposition on the printers' agenda.

UNIFORM COSTING

The Institute of Cost and Works Accountants (ICWA) defined uniform costing as: "The use of the same costing methods for different factories in the same industry or combine" (ICWA, 1944).

Uniform costing, therefore, represents an approach to achieving comparability of cost information within a single industrial sector. Solomons (1950b) suggested that much of the potential value of a uniform costing scheme lay in the improvements which it brought to the analysis of resource consumption within the organisation, particularly the identification of areas of waste and inefficiency where cost savings could be made. This benefit might be facilitated by the capability of uniform costing to reveal to each firm how their own costs compared with those of the best units in the industry. Unfortunately there is little evidence to suggest that this attribute was exploited by users of uniform costing. Indeed it was primarily to its role in pricing, either in assisting price maintenance or in justifying prices for Government supply (particularly in wartime), that uniform costing can attribute its popularity during the first half of the twentieth century.

The substance of uniform costing systems was prescriptive guidance on the operation of the system and the codification of the procedures and records involved in measuring output costs. Thus most uniform costing systems were based on practical manuals which were distributed to interested firms. A number of matters were usually addressed in the costing manuals:

> 1. The type and form of the source documents and record books to be kept. For example, stock books, wage records, job cards.
> 2. The procedures/structures to be adopted in establishing the costing system. For example, the categorisation of cost elements, the apportionment of overheads, absorption bases.

3. Practical guidance on the methods of computation, particularly where the potential for flexibility existed. For example, what costs to include in the system (depreciation, interest on capital), the rates to apply in computing these costs, on the treatment of capacity measurement and joint cost separation.

Through such instruction, the proper and complete application of the uniform costing system would provide a standardisation which would ensure comparability of the content, computation and presentation of cost data across the firms in an industry.

Had the cost analysis and inter-firm comparison possibilities of uniform costing systems been a more prominent reason for their adoption, the longevity of the schemes may have been greater. The advocacy of uniform costing systems as a way of 'beating the market' provided a more problematical objective, particularly in the context of cyclical economic conditions and in a political environment where anti-competitive behaviour was increasingly frowned upon (Walker and Mitchell, 1996). These conditions encouraged the pursuit of individualistic economic strategies which impeded progress towards uniformity. As was stated in an ICWA book on uniform costing which was published in 1944:

> There is no royal road to the establishment of uniform costing system. Patient, painstaking efforts are required, involving a whole-hearted, open-minded, co-operative attitude. The interest of the industry rather than that of the individual units, or members, must prevail. If decisions are reached, based on the well-being of the majority, success will unquestionably follow (p. 18).

Although it is no longer in vogue, the uniform costing movement provides a range of illuminating insights to a variety of important technical issues which have been addressed by the accounting academy. Most (1961, pp. 40-48; 1977, p. 64) has suggested that the standardisation implicit in uniform costing is a relevant precursor to the development of GAAP, financial accounting standards and inter-firm comparison schemes.

Uniform costing was also a significant influence on the development of overhead costing methods (Wells, 1978, pp. 68-69) and was viewed as a component of the scientific cost management movement of the inter-war years (Banyard, 1985, p. 23). Uniform costing also illustrates the potential for supra-institutional level influence on the accounting practices of firms and its dissemination demonstrates how accounting practice does not necessarily "sell itself" but has to gain adoption through activating techniques of persuasion (Napier and Carnegie, 1996, p. 6; Walker and Mitchell, 1996). Finally, uniform costing enhanced technical expertise in the costing discipline and generated skilled individuals who were influential in the professionalisation of cost accounting (Loft, 1986, 1988; Mitchell and Walker, 1995).

Trade Associations and Uniform Costing in the British Printing Industry, 1900–1963

THE EMERGENCE OF A COSTING PANACEA

The material reproduced in this chapter illustrates the economic adversities which confronted employer printers in Britain during the late nineteenth and early twentieth centuries and the deficiencies which were perceived in contemporary costing practice. The extracts from trade journals reveal the "insane competition" which pervaded the industry and which disadvantaged master printers in their relations with customers and labour. The way in which the organisation of employers was considered to be an essential precursor to any concerted attempt to alter the condition of the trade is also highlighted.

The emergence of costing as a possible solution to the printers' malady is tracked in early papers such as A.B.C. of Costing (1903), and extracts from Federation publications (Profit for Printers (1904), Printers' Costs (1909) and The Printer's Standard Price List (1909)). The latter were the product of the work of a Federation Costing Committee which was instituted in 1908. During the second decade of the twentieth century it became increasingly apparent to the BFMP that the prescriptions offered in these publications would only prove effective if they were applied universally by British printers. The need to devise a more widespread and enduring solution to the problems besetting the industry was made more urgent by circumstances which increased the cost of production. Legislation on National Insurance and factory conditions, together with increases in local taxation and the cost of type, necessitated advances in printers' prices if profit levels were to be maintained.

In 1911 a Committee of the Council of the BFMP was appointed to investigate the problem of costs and charges. In exploring the ways in which prices could be raised across the trade, the Committee studied the example of master printers in the USA. Mr R.A. Austen-Leigh of the committee was sent to America to investigate. The concept of securing uniformity in costing practice as a means of discouraging selling below cost had been successfully attempted on the other side of the Atlantic. Enthusiasm for costing had been kindled in the USA by convening mass 'Cost Congresses', the first of which had been held in Chicago in October 1909. The Committee on Costs and Charges resolved upon replicating the American experience in Britain.

HOW SHOULD A PRINTER KEEP HIS BOOKS?
(1873)

THROUGHOUT the whole region of Commerce, the importance of a simple, concise, and accurate system of book-keeping cannot well be overestimated, yet self-evident as this fact may appear, it is, like many other accepted truths, too often entirely disregarded or but imperfectly acted upon. Careless account-keeping is one of the primary causes of insolvency, and the reckless and absurd competition which has been especially conspicuous of late years in the Printing Trade, may, in a great measure, be traced to the imperfect knowledge many printers have possessed of the true and actual "working expenses" of carrying on their business. Printing is notoriously an anxious and exacting trade, making great demands on both the time and temper of the principals; it is one, moreover, abounding, in practical working details; the question, therefore, of how best to simplify and abridge the labour of keeping printers' accounts is one that merits serious consideration.

We have received from Mr, Bevan, accountant, of Sheffield, some specimen sheet of a set of books designed with this end in view, consisting of Printers Journal, Bought Journal, Cash Book, and Ledger. This gentleman has evidently had some experience with printers' accounts, and although we fail to discover anything strikingly novel in the forms sent to us, we are pleased to call attention to them as an effort in the right direction, and to express our approval of the simplicity by which their arrangement is characterized. We assume Mr. Bevan does not consider his set of books to be more than suggestive, as, although those here noticed might suffice for a very small jobbing office, they would prove quite inadequate to the requirements of a large and diversified printing business. We know of more than one office where much the same system as that advocated by Mr. Bevan is in operation, only with the modifications and additions proved to be necessary, with which it is scarcely fair to expect Mr. Bevan, as an accountant and not a practical printer, to be fully acquainted.

It would be impossible for us, within the limits of this article, to fully discuss in detail the various books it is desirable a printer should keep; we are, perforce, constrained to here confine ourselves to a few general remarks on the subject. In the first place, we are clearly of opinion that the method of "double entry," as intelligently and not slavishly applied to printers' accounts, is the only true and proper method of bookkeeping. . . .

. . . Reverting to the subject of book-keeping. A printer's profit is referrable entirely to his accurate knowledge of what is the real total cost of the job or work he is charging; the only way to ascertain this, is to keep a careful record of the various charges in each department through which the work passes, these charges ultimately finding their way, under proper headings, into what we may designate as a Work Composed (or Works Reprinted) Book and Job Book, or, to use Mr. Bevan's term, a Printer's Journal (his form of book with some modifications being a very good one). The system of charging job-work off-hand and from general experience, without reference to exact cost, is vicious in principle and as disadvantageous to the printer as it is unfair to the client.

(*The Lithographer*, 15 October 1873, pp. 63-64).

*

COUNT THE COST
(*1888*)

THE question, what does it cost to carry on a printing office? is one which every printer should be enabled to give a reasonable and definite answer to. His object in business is, of course, to obtain fair remuneration for his labour and capital, his expenses, his risks, and occasional losses. It is useful occasionally to suggest to him the question "Does he do so?"

The duty of the hour is to bring prices to a paying standard.

Most of the men who have become excessively rich, confessedly owe their great wealth to the unsparing use of printers' ink. Every enterprise that has put money into the pockets of its projectors and supporters, owes its success to the same means. No person seeks to employ the peculiar skill and knowledge of the printer without a full appreciation of the inestimable value of the facilities at the printers' command, and a consciousness that without them he cannot reach the public whose money he seeks to acquire.

What has been the printers' share of the fortunes he has helped to build? A drop in the bucket. And whose the fault? His own. Instead of establishing and maintaining the prices at which he was willing to take orders, he has permitted the customer to dictate to him the prices he was willing to pay. "The butcher, the baker, and candlestick maker," all retain the right to mark their goods, and if we do not pay the price of them we can go without. Why is not the printer just as firm in demanding a living price for his productions?

This matter has been one of serious concern to master printers for many years: In 1840, the celebrated Ambrose Firmin Didot, President of the Society of Printers in Paris, wrote an essay on the value of Composition. He said - "The master Printer whose charges are only fifty per cent. advance on the cost of labour, and who gives credit, is sure of inevitable ruin, which will be more or less speedy according to the capital he has to lose, or the credit that he will be able to command."

There never was more pressing need than at the present time for looking boldly in the face the question, what is the real profit - not the estimated or the apparent profit derived from an office-from the money sunk in plant, and from the energy and labour devoted to its management ?

In a business like printing, the greatest care should be taken to include even trifles at present almost constantly ignored. Unless this is done prices that seem remunerative will be found to be quite the opposite.

Much of the "Unprofitableness of Printing," so loudly complained of at the present time, is owing to overlooking the many small but in the aggregate heavy items of expense in every printing office. The principal of these are:- (1) Rent, Insurance, and Taxes. (2) Machinery and Standing Plant. (3) Type. (4) Brass Rule, Leads and Furniture. (5) Consumable Material. (6) Wages. (7) Goods. (8) Casual Expenses. In regard to machinery, the yearly deterioration is at least seven per cent. Leads and furniture, etc., depreciate at least twelve per cent.; rule twenty per cent. The consumable material includes coal, paper, ink, oil, roller composition, and gas. Other items explain themselves. A certain percentage ought to be laid on every job to meet its proportion of all these.

A prominent factor in working out a remedy is the question of the class of competition to which one is subjected.

Among the disadvantages incidental to the business is the multiplication of small badly furnished offices. Of necessity these compete from a vicious basis, by cheapening prices down to the level of their own inferiority.

That competition which depends for success upon the use of the cheapest and poorest material, and the skill of immature apprentices, debases the printer's vocation to a mere struggle for existence. It would be difficult to enumerate the many evils and disadvantages springing from this constant rivalry between the grovellers on the one hand and those on the other who would advance and elevate the art, and create a proper appreciation of it. It is equally difficult to discover an effective remedy; but it is better that the line of separation between the two classes should be made broader and more distinctive still by contrasts which cannot be overlooked. Therefore, let printers call to their assistance the best work of the type founder, the engraver, and the electrotyper; introduce, where needful, ornamental designs and graceful forms of relief to break the monotony of the page, study colour effects, and by all means improve their technical execution -this is the true line of progress, and, in the end, will defy competition from all but their peers.

(*The British Printer*, March-April, 1888, p. 12).

*

ENQUIRY OF REAL COST OF COMPOSING DEPARTMENT
(*The Printing and Allied Trades Association, London, 1892*)

A Sub-Committee have been making enquiry into this most important subject, and it is thought that the conclusions they have so far been able to arrive at may be interesting to the members of the Association, and may also induce those who have not already done so to furnish information for the assistance of the Committee.

The Committee have found it to be generally accepted by all who have given thought to the question, that under the prices now ruling, composing departments are actually worked at a loss, and the returns furnished in response to the request sent out by the Committee point conclusively to the true reason for this unsatisfactory condition of affairs. From these returns and from other materials placed at the disposal of the Sub-Committee it has been possible to ascertain with some degree of accuracy what percentage should be put upon compositors' wages to cover the direct and indirect charges of the department; and the Committee feel convinced that it is very far from the general practice to allow anything approaching a sufficient percentage when estimating for, or charging out, work.

A careful consideration of the returns shows that the percentage to cover direct charges only, such as reading and reading boys, overseers, proof-pulling and proof-paper, storekeeping and clearing, cleaning, &c., furniture and brass rule, depreciation of type, gas, and sundry departmental expenses, should in no case be less than 40 per cent., and that in houses doing much book-work it should be 45 per cent. or more. A very heavy percentage will even then be required to cover indirect charges, such as rent, rates and taxes, book-keeping, management, postages, discount to customers, commissions to travellers, bad debts, law and accountant's charges, and sundry small expenses; and although it is difficult to estimate the allowance for these charges, it is almost impossible that they can be covered by less than 35 per cent. to 40 per cent. The total percentage therefore necessary to cover all charges amounts to at least 75 per cent., exclusive of an allowance for interest on capital, profit, slack seasons, unchargeable overtime, &c. It should be remembered also that in houses where a scale lower than that of the London Society of Compositors is paid, the percentage would require to be even higher, as the direct and indirect charges are probably the same in these houses as in those paying the London scale; while in country houses the cost of carriage, &c., has also to be taken into account.

Further, in the case of bookwork the ever-increasing competition obliges printers to meet customers' requirements by getting up in type a large quantity of matter and by keeping it in type for an indefinite time, whilst in many cases the type is required to be practically new. These demands involve a constant outlay for type to an altogether undue extent as compared with the work to be done. It is frequently required, moreover, that long numbers shall be printed on paper supplied by the customer and often quite unsuitable for the purpose, leaving the printer the choice of ruining his type or bearing the expense of stereotyping. For all this increasing outlay, involving interest on capital and a large sum for depreciation, too often no charge can be made.

The recent increase in compositors' wages has not, your Committee believe, materially operated towards lowering the percentage necessary to cover direct and indirect charges, as these have themselves increased of late years and are still increasing.

It is notorious that under stress of unhealthy competition houses undertake work involving a large amount of composition at prices which allow but the slightest percentage on compositors' wages, with the idea of merely covering the cost of composition, and of making a profit in other departments. When this is done it is probably not realised how high the percentage should be to cover cost; and the Committee believe that in very many cases the result is a loss on the composition which the profit on other departments is not sufficient to cover.

It is difficult in the rush of modern business to find time or opportunity for careful investigation of figures and cost but the Committee would strongly urge the Trade to ascertain for themselves in their own composing departments how far the Committee's conclusions are correct, and having done so, to apply themselves resolutely to remedy such an unsound condition of affairs.

The subject is one of so great importance that your Committee are reluctant to consider their labours at an end and propose to give further attention to the question in the hope that they may be assisted by suggestions and communications from all members who have time and opportunity for enquiring into the matter.

(London Members Circular, Vol. 1, 1892).

*

THE FEDERATION OF MASTER PRINTERS AND ALLIED TRADES OF THE UNITED KINGDOM OF GREAT BRITAIN AND IRELAND
(1901)

The following important document has been issued, showing the proposed policy of the Federation:-
AXIOMS:-

1. THE ORGANISATION OF A TRADE, AND THE COLLECTION AND DISTRIBUTION OF INFORMATION TO ITS MEMBERS IS AN ABSOLUTE NECESSITY.

2. THE CONSTANT ADVANCE IN PRINTING METHODS RENDERS SUCH ORGANISATION MOST DESIRABLE FOR PRINTERS.

3. THE TRADE HAS HITHERTO BEEN ONE OF THE WORST ORGANISED.

Basing their action upon these axioms, it has been felt desirable to point out to those engaged in this important trade the need for them to unite.

METHOD PROPOSED:-

The system which offers most advantages is:-

1. For the Printers of each Town, or of two or three neighbouring Towns, to form a Master Printers' and Allied Trades' Association to watch over local interests.

2. For the whole of such Associations in the country to be Federated-so that matters common to all may be dealt with, while the provision of a centre affords facilities for gathering and disseminating information, and strengthens the component local Associations.

Local Associations already exist in London, Manchester, Glasgow, Edinburgh, Dublin, Leeds, Belfast, Bradford, Newcastle, Sheffield, Reading, Leicester, Halifax, Derby, Birmingham, Oldham, Norwich and Burnley, but some which are weak require strengthening and revived interest, while in places without Associations endeavours should be made to establish them.

The subscription to the Federation will, it is proposed, be paid by the local Associations, which will make arrangements to collect the necessary funds by a levy or increase of their own subscriptions.

The present arrangement is as follows:-

The Entrance Fee to be paid by each Association to the Federation shall be £1. 1s. with an Annual Subscription of 1s. 6d. for each £500, or portion of £500, paid annually in wages by the Association's members, the minimum Annual Subscription being £1 1s. and the maximum 200 guineas being subscription on £1,400,000 annual wage bill.

Or, the alternative option, That Entrance Fee £1. 1s. and annual subscription 7s. 6d. for each 25 hands employed, with a minimum annual subscription of £1 1s. and a maximum of 200 guineas, being subscription on 14,000 hands.

In places where Associations do not yet exist, individual firms will be invited to join the Federation direct, the entrance fee and subscription being the same.

WHAT THE FEDERATION IS FOR:-Some may ask what work the Federation is intended to undertake, and, although that must depend upon circumstances, it may be stated in general terms that it will do anything and everything that may appear to be for the benefit of the trade as a whole.

It is obvious that many questions arise in printing office management, where a well-organised central Association could render considerable help. It is hardly possible for any firm, however large, to have upon its staff experts of wide experience upon the technical subjects which are every day becoming more complex. There is need for greater accuracy in estimating the cost of work; new machinery and appliances have to be adopted from different parts of the world; printers' responsibility in connection with the Employers' Liability Act, the Workmen's Compensation, the Factory Acts, and similar laws, require wide expert knowledge. The occasional analysis of materials would sometimes prevent serious disasters. The question of Fire Insurance, and the liability of printers relating thereto is a very perplexing one. The internal arrangements of Works, so as to secure economical and efficient production, combined with safeguarding the health and comfort of the employees, involve questions of the first importance. Bad debts can be avoided by an efficient system of enquiry, and, in fact, there is hardly any side of our complicated industry where it might not be possible to secure greater efficiency by all the various members of the trade contributing of their experience to a central office carried on for the common good.

At the outset it should be declared that the Federation is not intended as an instrument to injure trades' unions, but wherever proposals for modification of conditions of labour are to be considered, the organisation would carry out any necessary negotiations, and it is well known that the old proverb "Union is Strength" applies with great force to labour questions, and that the existence of a powerful organisation on the side of the employers will tend to that calm consideration of any question at issue, which is so necessary to maintain harmony between the employers and employed of any trade.

There is no novelty in such a combination of Associations, it has produced good results in several trades in this country, while, under the title of "The United Typothetae of America," a similar organisation has been managing the printers' affairs of the United States and Canada with profit and pleasure to all concerned.

SOCIAL INFLUENCES:-

It has already been stated that Americans take pleasure in their Typothetae, for it has most enjoyable social gatherings at its Annual Conferences, and these are not weak factors in the history of a trade, as personal knowledge of other members of the trade would often result in minimised friction in business matters.

The London Association has its Annual Dinner, and at the half year a Conversazione. Other Associations have similar or varied arrangements, and after the first General Meeting the Provincial delegates enjoyed the hospitality of the Londoners, and while it is impossible to forecast the future, it is probable that some permanent arrangement may be made for the convenience or pleasure of the Members of the Association.

It is important to note that this movement was initiated and fructified in the Provinces, and that the Provincial Members decided that, although it was necessary for the Office to be in London, the General and Committee Meetings should be held at such Towns as from time to time appeared to be desirable. The Annual General Meeting for 1902 has been fixed for Manchester.

RESUME:-

In conclusion you are reminded that:-

You should, if not already a member, join a local Association.

That, if there is no local Association, or it is weak, a meeting should be called to revive or create one. A representative of the Federation will attend by appointment.

That, if not already done, you should press your local Association to join the Federation, or, as an alternative, do so yourself.

That each Association should do its best to encourage the formation of similar bodies wherever possible, or by attending meetings, strengthen existing but weak Associations.

The Council have instructed me to send you these particulars, with copies of rules and particulars of the General Meeting lately held, and trust that they will have your hearty support in their efforts for mutual good.

H.VANE STOW,
Secretary.
(*Members' Circular*, June 1901, pp. 61-64).

*

CHAIRMAN'S STATEMENT
(Annual General Meeting of the Federation of Master Printers, 1902)

The Chairman proceeded: The next item on the agenda is the question of unhealthy competition, and how it can be avoided. It is the custom with associations like this for the president to inflict upon its members an address. I am going to venture to speak to you chiefly upon unhealthy competition and its remedy. I hope that as a result of the reporting of this meeting the master printers of the kingdom may learn that the Federation is a living force, which has come forward to do good for the trade.

A new event has occurred in the history of printing in this country. An Association has been formed, with a long name and a short history, which aims at bringing together the whole of the employing printers of the United Kingdom. Its method of doing so is, in the first instance, to promote the establishment of local Associations which are federated to the central body. The Federation is yet in its infancy; what it will grow to depends upon the heartiness with which the various Master Printers throughout the country work together for their common good.

The subjects in connection with printing upon which one could speak are many. I need not take up your time by saying much to

THE IMPORTANCE OF THE INDUSTRY.- It is hardly necessary, at this time of day, to glorify the invention of printing. It is recognized as the maid-of-all-work to civilization, but, unlike the maids-of-all-work of the present day, it is badly paid. It is emphatically a progressive industry. In one important part of its work, the mechanical production of literature, it has met the public needs with a completeness which no other social agency has attained. It is not boastful to say that, at the present time, the methods of producing and distributing literature are so developed, that nearly all who care to read can obtain good literature - and, alas, inferior matter, if they prefer it - very often without cost. The difficulty is, not to find good books, but to find the temper of mind and the leisure to use them. Regarded merely as a manufacturing business, printing is an important industry. For example, there are nearly 100,000 persons employed in the printing and allied trades in London alone, apart from the immense number engaged in different parts of the kingdom. The industry, regarded as the servant of the public, is fairly satisfactory, but when it is considered as a business career for competent, educated men, I am afraid we must admit with regret, that the career of a master printer offers few prizes and many blanks. The reasons for this are many. Among the more obvious are the isolated and selfcentred way in which each printing office tries to work out its own salvation, very often to its own destruction and that of its neighbours. There is a growing feeling that this system of isolation and mutual suspicion should cease, and that master printers should associate themselves for cooperation in various ways. I therefore propose to confine this brief address to the subject of the

WISE USES OF THE EMPLOYERS' ASSOCIATIONS.- I speak advisedly when I say the wise uses, because there are uses which it is possible to aim at which are more than can be wisely or rightly done. We live in an age of combinations, and some of the greatest of them, chiefly of American origin, may become, if they are not already so, abominable tyrannies, which must be controlled. There is little fear that associations of employing printers can become oppressors of the public or tyrants over their workmen. There is a middle course between the isolated individual master printer trying to do everything himself, and fighting single-handed against the world, and the all-powerful combinations to which I have just alluded. It is possible for associated effort to do much to promote the best interests of the industry, not only for the benefit of the employers, but indirectly for the workmen, and for the public.

LOCAL ASSOCIATIONS ALREADY ESTABLISHED.- Though this is the first year of a Federation for the United Kingdom, a considerable number of Master Printers' Associations have been in existence in various localities for some time, and they are doing good service. No less than twenty-nine of them are now at work, of which the greater part are members of the Federation, and we hope as a result of this meeting that the remainder will come into line. They are connected with London, Accrington and district, Ashton-under-Lyne, Belfast, Birmingham and the Midland Counties, Bolton, Bradford, Burnley and district, Darlington, Derby, Dublin, Edinburgh, Glasgow, and the West of Scotland, Halifax, Hull and district, Leicester, Liverpool, Manchester and Salford, Newcastle and Gateshead, Norwich, Reading, Stockport, Sunderland, the Three Towns, Yorkshire, and there is also the Linotype Users' Association, which applies to the whole of the country.

WHAT ASSOCIATIONS HAVE ALREADY DONE.- They commenced on a very modest scale, with a small subscription, which members will do well to increase as the work develops. They are beginning to teach printers the long-neglected lesson that cooperation is better than mutual destruction. If competitors merely come face to face, and know each other personally, one step is taken towards a better understanding. The fact that these Associations increase in number and in membership shows that they meet a need, but their activity at present is, I must candidly confess, mainly a day of small things. At first the natural reluctance of conservative Englishmen to start a new scheme to co-operate with people whom they had hitherto kept at arm's length, had to be lived down. Some societies seem to think that their function is only to meet the men when questions of wages arise, forgetting that if no workmen's questions existed there are many ways in which combined action would be of great service. Already the re-arrangements of wages and hours of the utmost importance have been made with the trade unions with infinitely less discussion and difficulty than if each individual employer negotiated for himself. Some societies have put forth tentative efforts towards a living scale of prices. Others, and particularly the London Association, have watched legislation affecting printers. The London Association long since made a careful inquiry into the cost of composing work, thereby enlightening many printers as to the finances of this most anxious and unprofitable department. A valuable summary of factory legislation issued from London, which has just been referred to by Mr. Vane Stow, is in the hands of every member. These are only a few of the many services already rendered, and they are only a foretaste of better things to come. It may be still asked why are employers' associations more needed now than they were in the past. The need has grown up in consequence of

THE ALTERED CONDITIONS OF THE INDUSTRY.- Formerly, printing offices were small, and were carried on by hand-labour, with a few simple appliances; now the machinery is so varied and so complex that it requires an expert to understand it. Recent legislation has thrown upon the employer considerable responsibility, which was unknown to his fathers. Though this legislation is for the good of the community, it often bears hardly upon the individual. The master printer of to-day should have more qualifications than it is reasonable to expect in one poor mortal. He ought to have a good general education; he should possess a knowledge of the handicraft obtained by practical work in the different branches of the business. He should not only have considerable knowledge of engineering, but he would do well to understand electrical science. He must be a good administrator, able to deal with all sorts and conditions of men. He needs considerable knowledge of such complicated legislation as the Factory Acts, the Employers' Liability and the Workmen's Compensation Acts, and other useful social legislation. He should study the requirements, whether reasonable or otherwise, of the trade unions. He should have taste and some literary knowledge. To all these qualifications he must add an insatiable thirst for working harder, and for longer hours, and with more anxious wear and tear, than most business men, and, as a rule, for a much smaller income.

THE ACTUAL MASTER PRINTER.- He has, perhaps, as a youth worked at case, but has very seldom practically worked at all the departments. He has had no exact and scientific training in the difficult duties he undertakes. He either works up a business, or joins an established one, but in either case there is seldom any accurate scale of cost, or proper standard of charge. He is, therefore, a prey to the inconsiderate person, who being impressed with the idea that printing must yield large profits, because printers' prices vary so unaccountably, thinks it his duty to keep the printer humble by low prices and hard conditions. If the master printer does not improve and increase his plant, he is left behind in the race; if he does do so, he has the horror of seeing valuable plant lying idle and ruining him with interest and depreciation. Under these circumstances, he is tempted to take work at the best price he can get. The buyer takes advantage of his difficulties, and too often dictates not only low prices, but unreasonable and humiliating conditions. I do not say that all master printers are in this plight, but I appeal to my hearers who know where the shoe pinches when I assert that too often this is the actual state of the case. While the harried and anxious master printer is speaking with the enemy in the gate, he has internal difficulties to contend with. The trend of factory and sanitary legislation, of trade union requirements, and of many other movements, all add to his anxiety, and diminish his chances of income. Trades unions wisely require a certain standard of wages and conditions. They know their own minds, and their members hold together for their mutual interests, and I commend them for so doing. They set, in some respects, an excellent example to employers as to their dealing with the public, but this example they are wonderfully slow to follow. Master printers will admit the truth of all I say, and will deplore the present condition of things, but without association they are powerless to remedy them. Each printing firm is played off against another by an exacting public, and is intentionally kept in the dark about the unequal struggle it is waging. Though a certain amount of competition is good, reckless and ignorant competition is a crying evil. Printers should cultivate sound opinions as to continuity of work. It is obvious that if one firm has done a particular class of work for some time, and has adapted its plant and trained its employees in a special direction, it is wasteful in the last degree to dislocate this office by sending the work elsewhere merely because another printer has put in a fractionally lower price. None are more unreasonable in this respect than some public bodies. Many of these bodies rightly insist upon the workmen employed by the contractors being paid the standard wage, with good conditions of employment, but when they periodically advertise for tenders and accept the lowest, they know nothing and care nothing for the parlous state of the contractor, who, after investing for years his own or other people's savings in printing plant, finds that his fortune consists of a lot of old metal and worn-out machinery, because his carefully organized connection has been seized by another hungry printer. And the last state of both these printers is worse than the first.

WHAT IS THE REMEDY.- One remedy lies in the wise use of association. If printers throughout the country form local associations, and these are all federated to the Association represented here to-day, many of the master printers' difficulties will disappear. I have referred to the need of legal, scientific, and other knowledge, which from time to time every master printer must possess, or he must suffer through his ignorance. The Association ought to provide the best qualified experts for the use of its members, men who would, in the course of time, know more of the questions particularly affecting printers than any ordinary members of their professions. By arrangement and discussion, it is possible to elaborate a standard of workable conditions, which, altogether apart from price, are essential to the proper conduct of our difficult business. I venture to assert that anyone who understood the inner workings of a printing office would so harry the printer that, no matter what was the estimate given for the work beforehand, he could make any job a loss instead of a profit to the unfortunate printer. No one suggests that the public who require printing wish to be more unreasonable than the customer of any ordinary business. Their exactions are the natural result of our weakness. So anxious are master printers for work; so determined are they that when they have secured one job at a profit they will not he happy till they neutralize it by securing another job at a loss, that they practically invite the public to provide them with work under such onerous and wearing conditions that no one is benefited thereby. Some work, from its nature, must be done with overtime, with pressure and anxiety to all concerned; therefore the public considers that all its whims as to time can be gratified at the cost of the printer. The reasonable conditions needed are so numerous that one cannot mention them all. Some are prominently before us. As a rule, no one should object to give a price beforehand, but proper time should be allowed to prepare a careful estimate, otherwise errors of calculation are allowed to pass. Estimates should be given upon a carefully prepared specification. For want of it, A., who prints second quality work, gives a low price, while B., whose work is of the first quality, and therefore more costly, gives a higher one, whereupon the customer determines to have B.'s good quality at A.'s low price, and, if he pits one printer against the other, he generally gets his way. If he shows a truculent demeanour in his negotiations the printer's knees smite together, and he instantly and humbly offers a further discount. All this sounds very humiliating. I do not say it applies to every firm, and I admit there are thoughtful and considerate people, not a few, who want printing done under fair conditions and at a fair price; but too often my description of the negotiation between the printer and the customer is a true one, as some of my hearers can testify. Now, if the printer I have described becomes a member of a well-ordered association, some of these difficulties can be conquered. The Federation, with the assistance of experts from the local associations, ought to compile a schedule of prices covering the chief items of printers' charges. No doubt such a price list would be very elaborate, and would take time and money for preparation, but when it was once issued to members it would be a most valuable guide to them in estimating. Gradually there should grow up a public opinion against charging less than the scale which the collective wisdom of the associations found to be the lowest that would leave a living margin. Time would fail me to refer to the conditions forced upon printers by unwholesome competition. For example, in lithographic work; the offering of an unlimited number of free designs, which may all be rejected with contumely, is an obvious evil; working overtime without any extra charge is another. Occasionally printers are expected to work under time penalties, a condition which no printer should submit to. Type is kept standing without charge for an uncertain time. Work is put in hand, and is delayed by the customer indefinitely for his convenience, but when he restarts it, it must be pushed forward in a busy time with night work at the printer's loss, for whose difficulties no consideration is shown. The printer is under a statutory obligation to put his imprint on his work; to do so is his right, as well as his duty. If he omits it, the customer is not bound to pay for the work. In spite of all this, the printer often foregoes his right, and his duty, and the credit which he would gain by having his imprint upon well-executed work. These are only a few of the ways in which we allow ourselves to be imposed upon for want of a proper standard of conditions.

ASSOCIATION MEANS INTERCHANGE OF IDEAS.- The master printer - forgive me for saying it - needs teaching. He can learn much by conference with his competitors. In many respects the best-trained master, who learned his business twenty years ago, needs to go through a new apprenticeship. The developments of machinery have been so rapid and so complicated in recent years that it is questionable whether more than a small percentage of employers really understand the machinery they use. It is not because they are unable or unwilling to learn, but the peculiar difficulties and anxieties of a printing office so absorb the attention of its heads that there is no time for careful investigation. We need to take a broader view of printing office accounts. How many offices take fill-up work at a low price; but how few understand what this means. If certain work, to fill up, is taken at a little beyond the bare outlay for wages and materials, merely to keep the place running, but without including its fair share of general expenses, then it follows that other work must bear twice its share. The question of general expenses is a complicated one. It includes everything beyond the actual cost of wages and materials, and no one, without careful investigation, knows how great a proportion this is. Under general expenses must come not only rent, rates and taxes, depreciation, clerks, travellers, and office expenses, but there should be included 5 per cent. on the capital invested, and a reasonable sum for the salary of the employer. It is obvious that no man will intentionally carry on business, bearing its heavy burdens, unless he has some reward in addition to the salary he could earn in the open market, and to the percentage on his capital which he could get by a non-speculative investment. I venture to assert that, if the charge for printing was so systematically divided that for every job there was put in one column a sum sufficient to completely cover all these general expenses, then it would be found that the balance would often not cover the bare outlay for wages and material. I have had the opportunity of seeing many printers' balance-sheets, and some very mournful documents indeed, and my experience is that for every £100 charged for printing, the wages bill should not come to more than £43 for the actual workers, if the printer wishes to earn a small salary for himself, and say 10 per cent. per annum on his capital. The difference between the wages paid and the charge made may seem a large one, but the cost of materials, and still more of the vast unknown extent of general expenses, is so great that without this margin ruin ultimately follows.

CHEAP PRINTING IS A DELUSION.- There is no service that less deserves to be cut down in charge than printing. A great deal of the work is of the utmost importance to the customer, out of all proportion to the amount that is charged. It is often vital to the carrying on of his affairs. I claim that we ought to be held most firmly to efficiency, both in quality, accuracy, punctuality, and so forth; but for this efficient work the printer should be adequately paid, but at present this is practically impossible.

RELATIONS WITH EMPLOYEES.- I have already complimented the workmen upon their efficient organizations. The natural result of their collective action is that employers should meet them collectively. This need not be done in any antagonistic spirit. It is for the advantage of both employers and employed, and also of the public, that each of the parties to the bargain should be properly remunerated, and that the industry should be on a stable footing. While associated employers can deal with workmen's associations more efficiently than when isolated, so they can also deal with those employers, who I hope are not many, who are injuring the workmen by paying sweating wages, and injuring their fellow-employers by unreasonably low prices. All these questions want the light and the free air of discussion and publicity.

I have spoken, perhaps, too much of our rights; let me say a word as to our duties. We may find factory and sanitary legislation adds to our difficulties, but we know that it is for the public good, and we ought to comply with these Acts in a reasonable spirit. We live in difficult times for employers, and we are in danger of forgetting that, however well our workmen may be organised in their Unions, there should be a bond between us, something nobler than merely cash payments. There will always be an opportunity for the sympathetic and broad-minded employer to sweeten and brighten the lives of his employees in a thousand unobtrusive ways which will suggest themselves to every thinking man. Modern printing offices are more sanitary and cheerful and well-arranged than some of the miserable shanties of the past, but they might be cleaner, brighter, more healthful and more satisfactory than they are. No legislation can take the place of a sense of personal duty, though it can help to quicken it.

Modern relations between employers and employed tend to separate them into two parties, distinct from each other, in every sense of the word. We can do something to bridge over that distance. While we recognise and respect the independence of attitude of the workman to-day, we can still cultivate personal and friendly relationships in many ways. I am glad that the old idea is dying that the proper place of the wage-earners was to "bless the squire and all his rich relations, and learn, with all poor folk, to keep their proper stations." Good feeling and friendship are quite consistent with each party to a bargain trying to make the best terms for himself. Speaking of employers generally, and not specially of employing printers, I do not think we fully realise our responsibilities in this direction. We have in our offices a number of youths to whom we have undertaken, by a solemn document, called an indenture, couched in mediaeval phraseology, to teach the art and mystery of a printer. The heads of large businesses cannot possibly give the time to personally instruct each apprentice, but they can and ought to see that all that is possible is done to help the youth to develop into a competent workman, with the possibilities of a wider future. Is it the fault of the employers or of the apprentices that so many of them are not anxious to make the most of their chances of secondary and technical instruction? The London Association successfully brought its influence to bear upon printers' technical classes, to provide teaching at the Polytechnics for younger apprentices during business hours. This may be a small matter, but it indicates one of the ways where associations may promote better technical teaching. Is it our fault, or the fault of the printers' engineers of England, that nearly every mechanical improvement in our varied arts has now to come from abroad? We need a succession of "cunning hands and subtle brains," or our enthusiasm for the Empire will be only empty sound. We must stimulate our young men to active and stirring interest in training themselves for the responsibilities of life. It would, perhaps, be a counsel of perfection to urge that, when the Employers' Association meets, it should discuss its duty to itself, and then its duty to its neighbour. We may not be wanting in good intentions; as employers, but so pressing are the claims upon time, and thought, and strength, of those who find the brains to administer printing offices, that there is too little time left for thinking of one's employees' interests. Here the Association might help the individual to do what he could not do alone.

NEED OF PATIENCE.- We cannot expect that the associations will do everything for us at once. We need to understand each other and our mutual difficulties better. We need to cultivate more self-respect. We must be willing to pay large enough contributions to enable these organizations to work efficiently. We must have good secretaries, able and experienced, who, at any rate for the large associations, should give their whole time to doing for their members what the members could not do so well for themselves. It is a difficult task, but not an impossible one, to weld together into one association the employer of ten and the employer of ten hundred persons.

(*Members' Circular*, June 1902, pp. 93-103).

*

STAND FIRM
(*1902*)

It is beyond all question that many a printer's failure is attributable to lack of firmness in dealing with his customers. Necessarily, when a man of business buys, he is on the alert to do so in the cheapest market, that he in turn may be able to supply his customers at bottom prices. But printing being, usually, only an expense, he views it with a keener eye than, perhaps, any other of his purchases. Knowing also the great variation in printer's prices he is prone to believe, or to affect to believe, that whoever he is dealing with is the man who charges more than anyone else, and he thinks that by the process of squeezing he will be able to materially reduce his bill.

Even where a printer has been on good terms with his customer for a number of years, some little thing may arise which will lead to the impression that there has been an over-charge, and the printer is informed that, if he wishes to retain the orders, he must considerably modify his prices. Such cases are common, and naturally a feeling of anxiety is begotten lest a customer should be lost. Instead of immediately sitting down to reduce his charges out of sheer fear, however, a wise man will look into the matter, and communicate with his customers in a courteous manner, and ask them to point out the items considered high; as oftentimes with such firms it is the result of oversight in comparing the prices. Such a case recently came before the writer. For many years a printer had worked for a rising firm, whose account grew larger each year, and he was fully aware that such a firm would need to be served on the best terms, as, their goods coming largely before the public, they would be the objects of "attack" from printers' travellers in all directions. On one occasion an estimate was requested for a booklet of 10,000 copies, got up in good style, for which the price of £50 was given. Imagine the printer's feelings when he was informed that another house had quoted £32 10s. The work had been carefully figured, but was again examined, and the conclusion arrived at that the customers had made a mistake, as the principal item was the paper. This conclusion was communicated to them with a request that they would further look into the matter. The reply was that no mistake had been made, and this was accompanied by an intimation that they feared they had been previously over-charged. This was a serious matter, and one which called for judicious handling, as they were customers too valuable to lose if anything could be done to retain them. The opinion of some fellow-printer was obtained, one of whom quoted the same price, viz. £50, and another a few pounds in advance. The customers were again written to. They were loth to enter further upon the matter, but after continued respectful requests they did so, and found the quantity they had asked an estimate for was wrong, the other estimate being for 5,000 instead of 10,000. That printer was more firmly established with them than ever, and many orders from the firm have since been filled.

There is another class of customer by whom the fearful printer is more likely to be overcome than by the genuine mistaken one, viz., the man of "bounce." He always knows; he either has been in the printing business himself, or he has a friend who has been, and whatever price may be given it is always too high, and if the work is desired it must be considerably reduced. He further remarks that he can get the work done at so much less, but (becoming compassionate towards the printer) he would rather not take it elsewhere. The nervous man is first frightened and then touched by the apparent tenderness, so he accepts the work, at a price which he finds when completed leaves no profit at all. On the other hand, had he stood his ground, he would very likely have obtained the order at his original quotation.

Such at least was the case only a few weeks ago, where a young man had recently taken the reins of government in a firm who gave out large orders. An estimate had been obtained amounting to a considerable sum, whereupon peremptorily demanded the autocrat, "Why do you charge us fifty per cent. more than we can get this done by others ?" This was rather startling, as previously satisfaction had been expressed by the former chief. The answer was that the charge was a reasonable one and could not be reduced. Strange to say, the order was secured. Yet another recent example that came before the writer: A large quantity of circulars had been printed as estimated, and the customer wrote that he would have another similar supply if they were done at a much lower rate, otherwise he would not. The reply was an expression of regret at inability to meet his wishes unless inferior material were used. In this case also the order followed at the original price. In both these instances it was simply "bounce," an attempt to frighten the printer that he would lose the order.

It becomes increasingly important that there should be great carefulness in estimating; but having arrived at what is considered the cost, no amount of brow-beating should move a man from his ground.

The weak man is easily recognised by the "cutter," and becomes a ready prey to his methods. Nor is his trouble always ended when he has come down to the price that is demanded, for frequently, when the account is paid, a discount of five per cent. is taken, to which a feeble protest may be made, but finally allowed because "the cash will be handy."

It behoves all to treat their customers with the utmost courtesy, and to execute their orders with expedition; but, depend upon it, the man who is always ready to reduce is not the one who commands respect, for by straightforward men of business he is treated with suspicion; as they wonder, if he so readily alters his figures, whether he does not seek to overcharge at the beginning. Instead of increasing his business it decreases, and what he gets realises so little profit that he finds himself in difficulties, and is finally compelled to abandon his position as a master printer because he failed to stand firm.

(*Members' Circular*, September 1902, pp. 184-186).

*

THE GENTLE ART OF PRICING
(*1903*)

Some time back a printer undertook a job for 15s. He thought that later he might be rewarded with more profitable work if he continued industrious. After paying the actual costs of labour and paper, he was left with 1s. 3d., and an unbounded horizon.

He was a bit crestfallen when he was asked, some weeks later, to give another estimate for the same job, but he gave it, at the same price. He expected the customer would be in a hurry to clinch the bargain, lest the printer might be inclined to raise his price; but this modest son of Caxton had not reckoned with the other printer.

After waiting a suitable time, he enquired, as unconcernedly as he could, when he might expect the copy: to find that the commission had been given out at 12s. 6d., "less 12 per cent. discount for cash." The "successful" competitor has now an extra official to look after his vanishing interests; and the customer has secured an invoice that will be quoted against printers while there are any left who fail to detect the insufficiency of the figures.

Another printer was asked to estimate for the reprint of a job then standing in his office. The composition was worth £4 10s., and the other work was small. He decided to charge the composition at £3, to make sure of the job; but he did not get it. Now he is anxious to know how a printer can compose a job worth at least £4 10s. for something less than £3. When he has time, he wonders at the state of mind of a customer who finds that he has paid 50 per cent. more for a job than he need have done, if he had known the other printer earlier. The latter has not, up to the time of going to press, obtained the distinction of the under-cutter in the first case, but he bids fair to fail for a large amount if he is allowed to "estimate" long enough.

We have selected these cases because composition is the main item in both; and the differences in the price must have arisen from a failure to appreciate properly its selling value.

It is a suggestive fact that, while the art of pricing work has been fiddled with in our technological classes, we have not yet discovered a single model estimate, worked out by teachers of technology, which has not utterly failed to provide a fair selling price for composition, the totals being always against the printer.

We do not wish to attach too much importance to technological examples, the printer having no particular reason for taking them seriously. We use them to show how deep-seated are the delusions that afflict the printer's charging department. No one supposes that he designedly understates the value of composition. It is ignorance, and not merely senseless competition, that is fast reducing the printing business to the least desirable of all investments. The printer deceives himself more often than he pleases his customer. The latter does not know that he has something particularly cheap until he finds a printer who knows what a fair price should be. As he usually finds the other kind, he is kept in a worried and dissatisfied state. When he seems to have reached the lowest price, and clinches the bargain, he finds afterwards that a price still lower would have satisfied a printer In the next street.

But all through the declension the printer looms large as a prince of calculators. If a customer had a real respect for true genius, he would pay extra for the privilege of watching the trader wrestle with an estimate. It does one good to see such close attention to nice calculations: or would, if we did not know that these finely graded figures have for their base a fictitious unit of value. Printers having not yet settled upon a selling price for composition, estimates that bear no resemblance whatever to actual value will be a recurring feature in their business life.

Our artful arithmetician is on the look-out tower when making false quotations, having an airy notion that the thing will right itself although he starts it in the wrong path. He is an alchemist, is the printer, and has the magic power of transmitting a loss into a profit by leaving out an item in the cost. He is troubled with "unproductives," because, while he cannot dispense with their services, he neglects to debit the convenience to his customers.

Patience on a monument obtains our commiseration because she cannot get away; but the printer, whose monument is a case room that does not pay, excites a livelier feeling, he being a willing victim. The ancient jest, "Once a printer always a printer," may be a compliment, but is usually the acknowledgment of a strained and painful position. The "click, click, click" of the compositor's stick may not be as soothing to the printer as the "proputty, proputty, proputty" of the nag's hoofs to his owner, but whose fault is that? Certainly not the compositor's, who having spent his young life in learning a business, has earned the right to live by it.- *The British Printer, Nov.-Dec. 1902.*

(*Members' Circular*, January 1903, pp. 28-30).

AN A.B.C. SYSTEM OF COST-KEEPING
(Paper delivered at the Annual General Meeting of The Federation of Master Printers, 1903)

Mr. Sidney Reid (Messrs. Andrew Reid & Co., Ltd., Newcastle-on-Tyne) was then called upon to read the paper that he had prepared under the above title. The complete set of books and forms connected with the system was displayed in the room. Mr. Reid said:-

I think that no apology is necessary for the issue of this paper, as I am convinced that if the trade thoroughly understood the real cost of production of each order executed, and of all kinds of printing work undertaken by them, some of the estimating and selling below cost would cease.

It is not in calculating the first wages and material cost in which most of the mistakes are made when estimating for work, but in not allowing for the departmental and general working expenses of the business. It must be remembered that the printing trade cannot be carried on without a continual outlay in securing the orders, placing in hands of workmen, superintending, checking, packing, and delivering the work to the customers.

The system which I am now placing before you has been in use for the last twenty years, and can be thoroughly recommended. It has been developed and improved from time to time. The initial and annual cost is, and can be, kept very low, and many of the forms which I am now about to describe may be dispensed with for small offices, but the principal ones, such as the Work Ticket (Form No. 1), the Order Book (Form No. 2), and the Workmen's Daily Time Docket (Form No. 4), are as essential and necessary to the master printer as the cash to carry on the business. For 250 hands employed, an intelligent lady clerk and her assistant can do all the cost-keeping proper; consequently, the expense is small and quite within the reach of all printers. An important feature in this system is that the cost of any job in progress can be seen and examined day by day if the Workmen's Daily Time Dockets for the previous day have been entered into the Cost Book.

When introducing this new system, I should advise you to begin with the smaller departments first, and then follow with the larger ones.

The estimating and charging of all work should be kept in competent and trustworthy hands.

Those gentlemen who are in charge of large establishments can fully appreciate the value of a good, reliable system of cost-keeping.

I have made my explanations and examples as brief as possible, but I shall be very pleased, with the consent of the President, to give verbally any further details and particulars.

DISCUSSION ON COSTING SYSTEMS.- Mr. C. H. Lea (Messrs. Wertheimer, Lea &, Co., London) opened the discussion upon the foregoing paper as follows:-

In approaching this subject of costs, I venture to assume that no man will willingly and knowingly sell for 19s. what has cost him 20s. to manufacture, unless, of course, he has some ulterior motive, yet I am afraid that a very large percentage of the printing that is turned out every year in this country is done at a loss. It follows, therefore, that there is something radically wrong in the system upon which most printing offices are conducted, and that, as a consequence, the correct cost of the work produced is not clearly shown. Although we may well be ashamed to own it, this means-inasmuch as printers do not manu-facture until they have a distinct order-that we, as a class, are either too lazy, too careless, or have not brains enough, to ascertain what our manufactures cost us to produce. This, gentlemen, is the real cause of the ruinous prices and under-estimating, and I submit it is of the first importance that the Master Printers' Association should take this matter up and deal with it thoroughly, and not leave it until they have evolved the simplest and most effective system of bookkeeping and ascertaining costs that it is possible to devise for a printing office. So far as I know, no serious and systematic attempt has ever been made in this direction, and I am told that in America, as in England, printers have not arrived at a satisfactory conclusion on this subject. Yet it must surely be possible, by gathering the best points from the various systems in operation, to arrive at a practically perfect scheme. This is the first step towards putting the industry on a sound basis. It should be remembered that the printing business is two-fold. It comprises the work of the ordinary merchant or trader, and of the manufacturer as well; hence it is doubly important that the printer should have the best possible system of ascertaining costs and keeping his books that can possibly be devised. Many printers, like Mr. Reid, whose paper we have just listened to with great interest, have systems with which they are fairly satisfied, but the question is not are you satisfied, but is the system you use the very best that human ingenuity can devise ? I have searched the technical guides and journals in both England and America; I have lost no opportunity of inquiring into the various systems adopted by different printers with whom I have come into contact; yet I have sought in vain for a satisfactory system, and I am sorry to tell Mr. Reid I see many defects in his, some of which I shall presently point out.

Now I venture to define a perfect system as one that, while involving the smallest amount of work, will, in the simplest and most accurate form possible, enable the exact cost of all work to be readily ascertained, and will show the management clearly how each department is working; and, periodically, how each department, and the business as a whole, is paying.

Such a system should enable a man of ordinary business intelligence, whether a practical printer or not, to conduct a printing business successfully. And it should be remembered that this matter has a very important bearing on the value of a business, especially in the event of the death of a proprietor or partner.

I venture further to suggest that a sound system must at least comprise the following essential features:-

(1) It must enable the proprietor to readily see how each department is working, and how the business, as a whole, is paying.

(2) It must leave no opening for dishonesty, either by workmen, managers or clerks. With regard to this, if you will forgive me for straying for one moment, I should like to observe that I think a very heavy responsibility rests upon employers, who tempt men to dishonesty by their slackness. I hope the time will come when it will be an offence not to give information with regard to robbery of this kind, and that the State will not only have to prosecute the robber, but also fine heavily the employer whose slackness or carelessness has encouraged such robbery. It would save the ruin of many lives. But coming back to my point.

(3) A sound system must show the correct percentage of departmental and working expenses, and ensure that they are correctly charged on the prime cost of each job, and that the whole, with material and purchases charged as costs to customers, balance the total wages paid and other outgoings.

(4) It must ensure that the master printer, and not the customer or workman, gets the benefit of standing matter, full cases, or of a job being done in less than ordinary time from any other cause.

(5) It must, as far as possible, ensure the prevention of "milking" on the part of the men, and show the employer, as well as the foreman, exactly how each man has been employed. It must not allow a man to take it easy or waste time because he has a "fat" job, or a part is standing, or cases happen to be full.

(6) It must record the exact number of reams and hours, day and night, that each machine has worked during each quarter or year, as only by such records is it possible to know the rate per hour or ream that machines really cost. In a well-conducted office there should not be a startling variation year by year, if the overseer is careful not to have more cats than there are mice to catch.

(7) It must be inexpensive and simple in operation, and must, as far as possible, relieve the workmen of clerical work of all kinds.

These points, I think, all show the urgent necessity of this matter being adequately dealt with, and, if I may be allowed to do so, I will close by submitting the following resolution:- "That this meeting, recognising the urgent necessity for in improvement in the system of bookkeeping and ascertaining of costs generally in use in the printing - trade, urges the Master Printers' Association of the United Kingdom and Ireland to give the matter their careful consideration, with a view to placing before their members and the trade generally the best system that can possibly be devised."

Mr. W. Ashton (Southport), in seconding the resolution, stated that in his district eighteen master printers and master binders worked on the same estimate form, and they were now tackling the question of the best system of keeping the work tickets, and of bookkeeping generally. He thought that Mr. Reid's system was complicated, and would involve too much book-keeping, and suggested that, in any scheme that might be drawn up, due regard should be paid to the requirements and conditions of the smaller offices. (Hear, hear.)

Mr. W. Carter (Glasgow) thought that printers might assist in this matter by sending their suggestions for publication in the Monthly Circular of the Federation.

Mr. H. J. Waterlow (London) was curious to know how many jobs per day Mr. Reid could put through with the aid of his lady clerk and her assistant. In his business he feared that one lady clerk would not be sufficient, and if the superintendent of the composing room had to stay, as suggested, every night to check over the work tickets, of which there were from 2,000 to 2,500 weekly, he imagined that he would be there all night, and still not get half way through the work.

Mr. Walter Hazell (London) said it was most desirable that there should be some uniform method and basis for costing, and he thought the Federation should engage technical assistance to enable them to arrive at some definite idea as to what was, for example, the average proportion of general expenses to wages charged. If a number of representative firms would permit a chartered accountant, employed by the Federation, to extract such figures from their actual working as would enable a fair expense percentage to be arrived at, the Federation would be able to draw up a number of costing forms and account books bearing their stamp, and the use of such official papers in a printing office would do neither employers nor their workpeople any harm. (Hear, hear.) He believed this question to be one of the utmost possible importance.

Mr. Reid, in replying to the discussion, said that he passed about a thousand jobs weekly through his works, and his system was to put a foreman over every twenty men, and he found that these men could with ease check the work tickets for which they were severally responsible every night; and that one lady clerk and her assistant could easily carry the figures into the proper books. They had only two male clerks-the rest of the office staff consisted of lady clerks, and their work was perfectly satisfactory.

Mr. Lea's proposition having been carried,

Mr. W. B. Brewster, of New York, who is engaged in organizing work in connection with the Federation, delivered an address upon the subject: "What should a Printer Charge ?" His brief answer to that question would be: "Sufficient to enable him to pay 20s. in the £, to live as other manufacturers lived, and to have some margin over for necessary pleasure and recreation." The great thing to aim at was to put a stop, as soon as possible, to the suicidal competition that so largely prevailed, and he suggested some such plan as the following: Let the Master Printers' Association in any given district agree amongst themselves that they would not "cut" estimates merely to get the work away from a brother trader. There must he a fair price below which they ought not to compete against each other. When an estimate was being asked for, why should not the different prices proposed to be quoted by members be communicated, by telephone or otherwise, to the office of the local Association, and a minimum be mutually agreed upon? The members need not send in the same figure, but could submit any price they liked above the minimum, and the work would go to the member whom the customer selected. The members only needed mutual confidence in each other and mutual fairplay, and they would find that in the long run each would get his fair share, and at a better price than formerly. The dishonest customer who quoted imaginary lower estimates in order to bring down prices would be detected, and when members knew they were in competition with an unfair outsider, they could cut their estimate down to cost, if need be (but never below), and they would know, if they failed to secure the order, that their selfish opponent had undertaken the job at a loss. He added that in America the principle that he had been advocating had taken firm root, and the bulk of the printers in the chief cities were allied for mutual protection against the reckless price-cutter and the dishonest customer on the lines indicated.

Mr. Herbert J. Waterlow (London) remarked that the one great thing was to stand together and have confidence in each other. There were then many ways in which the Association could be of use to them. If, for example, they found that a customer was trying to cut prices, a communication could be made to the local association, and care would be taken that no member quoted below what might be considered to yield a fair profit on the actual cost of the job. This alone would be found to be advantageous. Another effect would be to check unnecessary competition of town with town.

A few words from an allied trader from Dundee, who said that the United Kingdom Billposters' Association had proved the great advantage of combination, brought an interesting discussion to a close.

(*Members' Circular*, April 1903, pp. 108-113).

*

PROFIT FOR PRINTERS: OR WHAT IS "COST?"
(1904)

To the Master Printers of the United Kingdom

THE extraordinary diversity of opinion as to the value of an average printing job, as evinced by several different estimates in competition, has always been a cause of surprise to consumers and printers; and, it is undoubtedly due to this wild estimating that the modern practice of putting almost every job up to competition has become so general.

It was with the desire to formulate a basis whereupon every printer might ascertain his prime costs, ensure some uniformity of price, and so check the tendency to "cut" prices on the part of the printer - and at the same time limit to some extent the tendency on the part of the customer to seek estimates - that a Committee of prominent London Master Printers undertook the difficult task.

The results arrived at are, it is believed, as exact as the intricate nature of the printing industry admits, and, in view of the divergent opinions as to the actual percentage of "dead charges" (*i.e.* wages and expenses that cannot possibly appear in an ordinary prime-cost sheet) which existed in the Committee, they found it desirable to retain the services of a Chartered Accountant to assist them in checking these by the figures actually appearing in the books of some representative houses of the trade, who were good enough to lay them open for his inspection, and his certificate is embodied in this Report.

"Prime-Costs."

These results have been found to fully confirm the opinion held by the Committee that the rates ruling in most printing offices as "prime-costs" are totally inadequate and misleading.

It is, unfortunately, a common error to believe that, if composition is paid for at the rate of ninepence per hour and the printer sells it at one shilling, he is getting a gross profit of 25 per cent. on his turnover, notwithstanding the fact that in many cases, he forgets to provide for the cost of the subsequent distribution. The figures which are given in the following report should satisfy him that a heavy loss must follow this short-sighted arithmetic. And to make up for the smallness of the *nett* profit which he hopes this will yield, he salves his conscience by persuading himself that "it will be made up for in some other department," where, generally, his calculations are equally at fault.

These results of ignorance or carelessness, together with a frequent cutting of the "fair" profit, and a not infrequent blunder in his estimate, all combine to *keep the printer poor, spoil the trade for his fellow printers, and bring a reputable craft into disrepute.* The customer is encouraged to place four or five firms in competition, knowing that one of them is sure to be well under his competitors, either by accident or design, and the printing industry, which - by reason of the laborious attention to detail that is needed, the high pressure, the risk of loss by spoilage, and the large amount of capital sunk in rapidly deteriorating plant and machinery - should be highly remunerative, has become a trade in which it is increasingly difficult to make a bare living, and whose members are constantly faced with the worry and anxiety of financial burdens which make them prematurely old and kill all legitimate ambition.

Look at your own Costs.

The Committee therefore earnestly ask for a close examination of the figures quoted here by every Master Printer in the light of the system of ascertaining prime-costs which prevails in his own office, believing that the absolute necessity for gradually raising prices must be brought home for him in order that he may reap a due reward for his labour. And it must be patent to everyone that, even if this course should, for a time, have a tendency to reduce his output, yet at the end of the year he will be better off, because he will have secured a larger profit on a smaller turnover, with a corresponding decrease of wear-and-tear, both to himself and his plant.

If a printer elects to pursue a course of haphazard estimating and pricing, and goes under, he cannot command much sympathy; the object of publishing the Committee's conclusions is to afford reliable information to the man who is anxious not to deliberately err in the conduct of his business, but who, through one cause or another, has not yet grasped the intricate principles which govern profitable factory management.

Prices not wilfully Cut.

Prices are not always wilfully cut; it so frequently happens that the printer believes he can see a profit, when, actually, there is a loss. In a great majority of cases a fair price can be as easily obtained as a low one; it only requires a little stiffening of his back for the printer to secure this. The day for "cheap" printing is passing; it now remains for the craft to mutually co-operate with the object of securing a healthy competition in *quality*, and they will find that the customer is awakening to, the fact that "the lowest price is not always the cheapest price," and that he will prefer to go to the firm who gives him the best service rather than to that which is chiefly concerned in finding out how *cheaply* the work can be done, instead of devoting its energy to discovering the best means of doing it well.

Others agree with us.

The Committee have been much encouraged in their labours by the signs of unanimity on the pressing need for a change in the present system of costs which are afforded by the meetings of Master Printers now being held in every part of the Kingdom.

They append the results of such meetings in Belfast, Glasgow, and Derby, and are glad to learn from reports received, and also from their own individual experience in London, that there is a distinct hardening of printers' prices, and that the movement which has been initiated for the betterment of the trade is already bearing good fruit.

They have a appended also:-

1. Part of the very interesting Paper, by Mr. W. W. Fox, on "The Costs of the Case Room," together with

2. A Report of the Conference which was subsequently held, at which a large number of Printers were present and spoke on the question.

3. The conclusion of a Paper on "Printing: a Fine Art and a Paying Business," by Mr. J. Benjamin Gotts, and

4. A Paper on " Machine Room Costs," by Mr. W. A. Guest, all of which have a special interest and bearing on the subject.

With the display of a more charitable feeling towards his competitors, and a careful regard for justice to himself in his dealings with his customers, the Committee firmly believe that a brighter time is in store for the printer, and that the day is not far distant when "the art preservative of all other arts," will cease to be a profitless vocation to him who practises it.

In order to show the need for such an investigation as the Committee have carried out, they invited fourteen London firms to tender for the printing of a well-known Trade Journal, each firm being provided with precisely the same specification. This table shows the diversity of "judgment" displayed by printers in preparing important estimates.

COMPOSITION.	CHARGE FOR STANDING MATTER.	MACHINING.	BINDING.	PAPER.	GRAND TOTAL.	REMARKS.
£ s. d.		£ s. d.	£ s. d.	£ s. d.	£ s. d.	
59 17 8	2/-	7 18 0	1 13 6	13 5 1	82 14 3	All Comp. taken at 1/3 per hour, with 25%, added, and same time estimated *for all text portion.* Imposing, 2 hours per sheet of 8 pp.
76 9 0	Allowance of 20 per cent. on Comp. prices.	16 13 0	3 3 0	13 15 0	110 0 0	Make up, 1/3.
72 3 9	1/3	9 18 6	1 11 0	13 8 6	97 4 8	Adverts.: Brevier, 16/-.
76 18 9	Allowance— Intact 4ths Remade up 4rds.	12 8 6	Included in Machining.	14 2 0	108 9 8	Lino price. Long Primer Tables.
78 15 0	3/-	8 5 0	2 3 6	13 7 0	102 10 6	
89 16 4	—	11 7 6	2 2 0	11 15 0	115 18 10	
128 3 6	2/6	8 6 9	2 12 6	11 11 0	150 13 9	Composition taken at 1/6 per hour.
68 13 0	2/-	11 4 6	2 2 0	13 5 0	95 1 0	Blocks to be charged at 3/8½.
76 13 0	2/6	6 9 6	1 18 6	12 0 0	97 1 0	Comp. of 4 pp. Suppl., £5 7 0
114 17 0	Rent 4d.	11 4 0	2 2 0	15 6 6	143 9 6	„ 4 pp. Cover, 4 12 0 Double. 2 cols., ¼ extra ; 3 cols., ⅓ extra. Make up, 1/- per page.
86 1 7	1/6	10 2 3	1 16 9	12 10 1	110 10 8	An "all-round price."
82 3 0	2/-	11 0 6	2 8 0	13 2 9	108 14 3	Solid. Leaded. Wrapping, Stamping and Posting, 7/6 per 1,000.
33 6 8	None	14 0 2	3 0 0	15 5 0	65 11 8	Imposing, &c., 16 pp., 11/6. Corrections included. Overtime extra. *Less* 7½% *discount.*
83 9 5	2/6	8 10 0	1 11 6	11 10 6	105 1 5	Making up, 16/- per sheet of 16 pp.

(Profit for Printers: or What is "Cost?", 1904, i-vi).

WHAT IS THE COST OF PRINTING IN LONDON ?
(1904)

REPORT OF THE LONDON COMMITTEE

THE CASE ROOM
A committee of London Master Printers, representing large and small houses doing every variety of Book, Magazine and Jobbing work have, after many and protracted meetings and the employment of a Chartered Accountant, arrived at the conclusion that, as a general principle, it may be taken that a compositor, from the time he enters your Case Room until he leaves it,

Costs on an average 1/6 for every hour.

The Basis.

In the following details, and in the Accountant's statement which accompanies them, *Wages* are taken as a basis. Higher paid men are set off against apprentices, so that 9d., the average wage paid per hour, represents the ordinary compositor at the 39/-. minimum for 52 1/2 hours.

In *Jobbing*, where the ordinary method of working out an estimate, or charging up,is to reckon so many hours composition, at say only 9d., on the compositor's time bill, the time occupied in distri-bution would then become an extra cost, equal to an extra quarter of an hour, BUT THIS IS FREQUENTLY FORGOTTEN.

What to add to find the Departmental Cost.

Having started with the 9d., the next thing is to go through the wages book, see what other hands are being paid in the department, and strike a percentage of this sum against the compositor's wages.

Reader, Storekeeper, Overseer and other Wage Charges.

Reading is the highest item, and the wages of readers and their boys will be found to average about 18.85 per cent. on the compositors wages. We have next to consider the clickers and, taking their wages out separately, the Accountant shows an average percentage against the compositor's wages of 9.25 per cent. Their work at case, however, varies according to circumstances and the size of the "ship," so we think the addition of 5 per cent. sufficient to cover their time not occupied in composition, instead of 9.25. The storekeepers and "clearing," apart from distribution, account for 4.29 per cent. We then have on the wages book errand boys, and, in a large office, a hand or so for odd jobs, who will represent 2.29 per cent. The expenses of the overseer and his clerk show a percentage of 4.89 per cent., and we must allow .50 per cent. for proof pulling-*not clean proofs*, for which an extra charge should always be made. We have, therefore, to deal with a percentage of 35.82 per cent. upon the 9d. per hour, making the *wages cost* of the composition 1s 0 1/4.d.

Depreciation and other Departmental Cost.

Before leaving the Case Room, we have, to deal with other expenses which do not appear on the wages bill. The depreciation of type should be allowed for on every job, but in practice it is dealt with very differently by various houses. Where type is much in use the depreciation is naturally greater than where a large quantity of work is kept in standing formes. Some houses treat this depreciation as an Establishment Charge, along with all other plant and machinery, and take of an annual percentage from their plant valuation of from 5 to 10 per. cent. If it is to be considered a Case Room charge, it appears to average 11.74 percent. on wages, and we recommend the adoption of this figure in ordinary jobbing houses, where it should enable the Master Printer to see his plant replenished, and give him a small margin to substitute new for obsolete founts. Depreciation of fixtures is 1.12 per cent. To this must be added the average for rent, rates, taxes, insurance, light, heat, repairs on the Case Room only, which shows 9.42. Other charges, such as proof paper, ink, type wash, brushes, mallets, &c., are covered by .27 per cent. These percentages amount to 22.55 on the 9d., or 2 1/8d., thus making a total time cost of 1s. 2 3/8d. per hour.

In our opinion, therefore, 1s. 2 3/8d. per hour is the prime cost to which to add your Establishment Charges (other than Departmental Costs) and your subsequent profit. The possible saving in some particular houses has not been overlooked, but many charges, equally variable, may be put on the other side as quite balancing it.

Establishment Charges.

Having arrived at a figure for the composition, we have, next to consider how the general expenses (Establishment Charges) *outside* the Case Room and other earning Departments are to be covered. These include:-

Rent	Heat	Cash Discounts
Rates	Counting House Salaries	Bad Debts
Taxes	Travellers	Repairs to Premises
Insurance	Order Office	Depreciation of Fixtures (other than Departmental)
Water	Stationery and Stamps	
Light	Interest on Capital at 5% p.a.	Cartage and Messengers

and a proportion of the total cost of these items must be charged to the cost of the Case Room and each other *earning* department.

After a careful examination of figures, and comparison by many printing establishments, these charges appear to represent a further 40 per cent. on the 9d. wages, which, when added to the Departmental cost of 1s. 2 3/8d., brings it up to 1s. 6d.

In our view, the only safe way, practically, of dealing with these Establishment Charges is to take out the annual amount and put the percentage against the wages paid during the same period. The actual analysis varies considerably, and it would be as well for each master to make his own. He will no doubt be frightened at the result but if he is not putting his composition at a cost which will cover all these items he is losing money in the Case Room.

Distribution.

We now come to the question of Distribution. We have taken all the above percentages on the 9d. paid to the compositor for one hour's work, but must again point out that, for every hour of work done by the compositor in setting up, about a quarter of an hour, on the average, must be added for *distribution,* so that the 1s. 6d. per hour should be charged for the time taken in setting up, plus 25 per cent. extra for distribution.

Thus, if a job takes 100 hours to compose, it will take 125 hours to compose and distribute. The cost of the job should be reckoned out at 1/6 per hour for 125 hours, not 100, i.e. £9 7s. 6d.

The above is very important, and not infrequently forgotten.

Of course, where work is done on piece distribution is included, and it will, therefore, be sufficient to add the percentages to the piece cost.

Conclusion.

The conclusion we arrive at is, therefore, that the average cost of a compositor, per hour, cannot be reckoned at less than 1s.6d.

We have tried to make the cost comprehensive, so that if, under the stress of undue competition, composition be charged at only 1/6 per hour, not only should there be no loss, but 5 per cent. will be earned on the capital employed. It is obvious, however, that no one would intentionally embark in the anxiety and uncertainty of business unless he expected his capital to earn considerably more than 5 per cent. per annum, as well as a living wage for himself, for which item no allowance has been made.

Our enquiry has been restricted to the cost of hand composition. We believe that experience shows that the nett saving of Machine Composition on *miscellaneous work* is small, after allowing for the interest on, and depreciation, of, the expensive and short-lived machinery, and for the liberal wages which the operators earn.

We recommend all Master Printers to make these figures a standard cost, for all work done in London, though some of it will cost more than the sums above mentioned. Bearing in mind the ill-informed competition of the past, it will, probably, not be possible to secure the profit on these costs immediately, but they are a standard to aim at for the future.

The Accountant's figures and certificate are as follows:-

		d.
Compositor's Wages	100.00	9.00
Readers	18.85	1.70
Clickers	9.25	.83
Storekeepers and Clearing	4.29	.39
Odd hands	2.29	.21
Overseers and Clerks	4.89	.44
Depreciation of type	11.74	1.05
Depreciation of fixtures	1,12	.10
Rent, Rates, &c.	9.42	.85
Proof paper, &c.	.27	.02
	162.12	14.59
Establishment charges	40.00	3.60
	202.12	18.19

"Having examined the figures of various returns furnished to me by houses of all classes, where wages paid to Compositors range from £30,000 p.a. in the largest to under £3,000 p.a. in the smallest, and having carefully dissected them, I beg to certify that the average cost of hand composition in these houses is 1/6 per hour.

(Signed) THOMAS L. THEOBALD."

57, MOORGATE STREET,   Chartered Accountant.
LONDON, E.C.,
April 9th, 1904.

THE MACHINE ROOM

We have gone very carefully into the costs of the Machine Room and find that, including as they do a heavy rental, driving power, a very heavy item for depreciation, and a large variety of other expenses, they will work out in the aggregate to a considerably higher figure than the costs of the Case Room; and that the Printer must on an average add something like 150 per cent. to his wages in order to cover his total expenses.

But in the Machine Room the expenses are not only in the aggregate a larger amount than the wages, but they depend entirely upon different circumstances, and bear no definite relation to the wages; so that we have come to the conclusion that it is necessary to calculate in some way the expenses *per hour* of each machine, or class of machine, instead of adding a percentage to the wages, and it is convenient to get a figure to include not only the Departmental Costs but also the Establishment Expenses.

What to Include In Cost.

The Printer should take out his total wages bill of the year, and his total of expenses, in which are included

Rollers	Rags
Turps	Oil
Blankets	Tape
Set-off	Repairs
Depreciation	Driving power
Rent	Rates
Taxes	Water
Repairs to building	Insurance
Light	Counting house and order office
Interest on capital	Bad debts
Discount	

If he gets the grand total of all these items, so far as they refer to the Machine Room, he can then proceed to divide them over each machine in proportion to its size and cost. Then again he can divide this sum which represents the year's expenses over the number of hours that the machine is employed during the year, and so get a cost per hour for each machine.

How many Hours is a Machine employed ?

This is a difficult question, which is answered differently by almost every house. What we want is to ascertain the number of hours during which the machine is either used for making-ready or for running, and is therefore chargeable to the customer. In every house there must be certain slack times, or intervals between jobs, which are not so chargeable, but on the other hand there are additional working hours in overtime. Allowing for holidays the whole number of ordinary working hours in the year amounts to about 2,700; and setting off overtime against a portion of the lost time we have taken 2,500 as the *maximum* during which a machine can be expected to be in use, and our figures are based on that assumption. Several houses, some large and some small, have proceeded to calculate the cost of each class of machine per hour on this basis, and the figures we give below are the average of their results. They represent therefore the dead cost of each machine per hour, whether making-ready or running, and are exclusive of ink, and the cost of the paper warehouse and of the printer's profit.

The printer therefore in using this scale, or one that he has prepared for himself in a like manner, should remember that he has also these three other items to add

Ink.   Warehouse.   Profit.

If the printer takes the trouble to investigate, he will doubtless find that his warehouse costs him a considerable amount in rent and wages. He probably recovers a share of that cost by profit on paper when he supplies it; but when, as is often the case, he is printing on paper supplied by his customer, he must charge either a percentage on the machine work or a separate item to cover this expense.

In arriving at the following prices for different classes of machines, the London Committee have adopted a system of units similar to, though not the same, as Mr. MacLehose in his paper (*vide* page 27).

Cost of working Machines, per hour (exclusive of Ink and Warehouse charges).

PLATEN.

From Foolscap Folio .	1/-
to	
Demy Broadside .	1/6

WHARFEDALES.

Up to Royal . .	2/-
Double Crown . .	2/3
Double Demy . .	2/6
Double Royal . .	2/9
Quad Crown . .	3/-
Quad Demy . .	3/3
Quad Royal .	3/6

TWO-REVOLUTION, (AMERICAN OR ENGLISH) & TWO-COLOUR WHARFEDALES.

Quad Crown . .	3/9
Quad Demy . .	4/-
Quad Royal . .	4/3

DOUBLE CYLINDER.

Quad Crown . .	4/6
Quad Demy . .	4/6
Quad Royal . .	5/-
Quad D. Foolscap .	5/-

N.B.—The make-ready, as well as the run, should be reckoned for at the above rates.

COMPARISON OF RESULTS.

In conclusion it will be interesting to compare the results arrived at by independent investigators as follows :—

CASE ROOM.

EITHER COMPOSITION OR DISTRIBUTION.

London. Per Hour.	Belfast. Per Hour.	Glasgow. Per Hour.	Derby. Per Hour.
1/6	1/6	1/6 to 1/7	1/7

MACHINE ROOM.

MAKING READY OR RUNNING.

	London. Per Hour.	Belfast. Per Hour.	Glasgow. Per Hour.	Derby. Per Hour.
Platen	1/- to 1/6	—	—	1/-
Wharfs. up to Royal }	2/-	—	—	—
Double Crown	2/3	2/3	—	1/7 to 2/-
Double Demy	2/6	2/6	2/1½	—
Double Royal	2/9	—	—	—
Quad Crown	3/-	3/-	2/10½	—
Quad Demy	3/3	3/6	3/3	—
Quad Royal	3/6	—	—	—

(Profit For Printers: or What is "Cost?", 1904, pp. 1-8).

PRINTERS COSTS
(*1909*)

What constitutes a profitable business? We ask this question in all good faith for many master printers are struggling-on contented so long as they are getting a living even though it be but a little better than that of one of their departmental foremen. As a matter of fact no business can be regarded as profitable unless it pays a salary to the owner as manager (if he acts as such), interest on the capital invested, and a profit on the year's working after *all* expenses (including insurance, depreciation, etc.), have been paid.

We are reminded of this by the arrival of an exceedingly useful pamphlet entitled "Printers' Costs," recently published by the Federation of Master Printers. It suggests a complete system (which can be adapted or modified to suit the special requirements of any printing office), of locating the jobs which do not pay and departments or individuals not fully employed or otherwise unremunerative. This system is not the work of theorists, neither is it a new way of juggling with figures. It is rather the combined work of five of the best known men in the trade to-day, men who are practical printers, men who are presiding over the destinies of some of the largest and most successful concerns in the country.

It cannot be too strongly emphasised that only on the solid rock of cost knowledge can any business be safely and permanently conducted. Once the actual cost is known, and not before, the profit can be added, and it will be a profit genuine and real. How few printers give the matter serious thought, preferring to grope along in the darkness of uncertainty, unable to understand how it is that although they have had a good year at apparently good prices they find themselves financially just where they were twelve months ago, having achieved for themselves no lasting good and done the trade positive harm.

To find out what any job has cost at any moment from its acceptance as an order to its delivery to the customer is a simple matter, and from the work under notice a system suitable to the needs of any office, no matter how unusual the character of the work, can be easily selected.

The pamphlet is not merely descriptive. It is accompanied by a complete set of specimen forms, nicely ruled and printed, and with various jobs charged up, forming at once the most lucid and thorough effort in that direction we have yet seen. All members of the Master Printers' Federation receive a copy gratis, but it has been compiled for general use, not with the idea of benefiting members alone, but of uplifting the trade as a whole, of putting it on a sure foundation, and of restoring to the printer the dignity attaching to an honourable profession.

Copies of the pamphlet (together with examples and specimen forms), can be obtained post free for 2s. 6d. from the offices of the Federation of Master Printers.

(Members' Circular, September 1909, pp. 181-183).

*

PRINTERS' COSTS
(1909)

PREFACE

A CONSIDERABLE amount of enthusiasm and self-sacrifice is necessary to induce such proverbially busy men as Master Printers to devote their time and thought to the production of a system of Costing and Book-keeping. Each has probably already decided upon the system peculiarly applicable to his own business, but the recommendation of a system suitable for both small and large houses involves reconsideration of all the points.

The gratifying reception accorded to the book already issued by the Federation, entitled "Profit for Printers," has, however, induced certain of our members to be good enough to undertake this task-not a thankless one, as we are sure that our Council and Members will greatly appreciate their labours: they are Messrs. W. Howard Hazell, A. F. Blades, Harry Cooke, C. H. Lea, and W. H. Ibbetson. His fellow-members wish special thanks to be accorded to Mr. Howard Hazell, who very kindly undertook the bulk of the work in preparing the pamphlet and forms, and was Chairman of the Committee.

The Committee trust that this pamphlet will prove useful, even although not universally suitable without modification, and they will greatly appreciate any friendly criticisms or suggestions for a future edition.

H. VANE STOW
E.TAYLOR THOMLINSON
Joint Secretaries

FOREWORD

THIS pamphlet on COST-KEEPING for Master Printers has been prepared by a committee of five of the members of the Master Printers' Federation of Great Britain. The members represent firms in London and the provinces, and it is the result of their deliberation, extending over several months.

It was found that each office had its own methods, and in this pamphlet the committee have endeavoured to take from each system the best points, and combine them in a system which is suitable, or readily adaptable, for any sized office.

A printer who adopts the system recommended may find that he will have to make various modifications to suit his own business. The essential points of a satisfactory Cost-Keeping system for printers, and points which are fully covered in this pamphlet, are as follows:

1. The business should be divided into departments.
2. Each department should be debited with all its expenses, and credited with the value of work done.
3. As far as possible, the time of every worker in the factory should be charged to the "job" on which he is working.

If this is carried out, jobs which do not pay are immediately located, and workers who are not fully employed are noticed, and the departments not paying are soon seen.

The system advocated does not interfere with the ordinary book-keeping system of a Printing Office. The only necessary additional book required is a small Departmental Ledger, which will last many years and take very little time to keep. The essential features of the system are:

1. Day Book and Bought Journal, ruled with analysis columns.
2. Productive workers to write a Daily Docket.
3. Work Ticket, upon which the time, cost, and materials used on each job are entered.
4. Departmental Ledger, containing the Departmental accounts, which may be posted up either monthly or quarterly. The Departmental Ledger will be distinct from the accounts used for making up the Balance Sheet and Trading account.

(*Printers' Costs*, 1909, pp. 2-4).

*

THE PRINTERS' STANDARD PRICE LIST
(*1909*)

INTRODUCTION

A PRICE LIST FOR PRINTERS! Impossible, is the obvious first criticism, when it is remembered that no two jobs are alike. And yet there are several such lists extant, which, from their sale, are evidently of service. The prices, however, are twenty years or more old, and conditions have been materially altered since that time.

Recognising these points, the Master Printers' and Allied Trades' Association felt that a real purpose could be served by the issue of a List of Prices, brought up to date, which would give some guidance to printers who are accustomed to deal with the miscellaneous work such as is comprised in the following pages. A Sub-Committee was therefore appointed, and much time has been spent in going into various details; and while the present book does not presume to be perfect, it is offered as a guide to a Standard which is worth striving for, if not always attainable.

Issued by the Association, it has an official character, and dissatisfied customers who are referred thereto will be more likely to accept a charge made, as not excessive. In this way the List should be most useful.

A word or two of explanation is necessary. It became evident at the outset that the prices charged for small quantities were utterly profitless. The guiding principle in the past, and recently advocated by a writer in one of the trade papers, seems to have been to "decrease the charge on small jobs in estimating and increase it on large ones." This the Committee considered to be fallacious. The prices for the very small quantities given herein may at first, sight appear startling, but they are based upon the costs as worked out in "Profit for Printers," and are such as should be obtained if the printer is to receive a fair remuneration.

(The Printers' Standard Price List, 1909, pp. 7-8).

*

COSTS AND CHARGES
(Annual General Meeting of The Federation of Master Printers, 1912)

The report of the Committee on Costs and Charges was then read as follows:-

REPORT OF THE COMMITTEE OF THE COUNCIL ON COSTS & CHARGES.

At the Council Meeting on October 10th 1911, after a long discussion on "Increasing cost and necessity for increased charges," Mr. J. E. T. Allen, Mr. W. A. Waterlow and Mr. R. A. Austen-Leigh were appointed a Committee to consider and report.

Feeling convinced that the best, if not the only, method of obtaining practical results in the desired direction was to approach the question on the costing side, and to secure the adoption of a uniform costing system, the Committee invited the co-operation of Mr. Cooke, Mr. Howard Hazell and Mr. C. H. Lea, who had been members of the Committee on Costing which, in 1909, compiled and issued "Printers' Cost, a System of Book-keeping with Examples and Specimen Forms," Mr. Howard Hazell having prepared and edited the copy. A meeting was held for preliminary discussion on October 31st, when the Committee adjourned until the middle of January, by which time Mr. Austen-Leigh would have returned from a visit he was about to pay to the United States, where he expected to have opportunities of learning more of the American system from the organisers and officials of the American Printers' Cost Commission.

Meetings were held on January 24th, January 31st, February 28th, May 2nd and May 9th, and advantage was taken of the Council meetings on December 5th and February 27th to discuss the matter with members of the Council. The Committee examined, as far as lay in their power, existing costing systems, especially the American (which has been adopted by some houses in this country), and what is known as the Westminster system, claimed to have been formulated before the American system, and in many respects identical with the American system. In the case of the Westminster system they had the advantage of receiving information and reports direct from the originators, Mr. Roberts (auditor to the Federation) and Mr. Walmsley, who kindly placed their services at the disposal of the committee. Mr. Roberts is present at the Annual Meeting to-day, and is prepared to give further information and assistance if desired.

The result of their examinations of these systems led them to believe that the principles enunciated in "Printers' Cost" are correct in the main, though the method of carrying these principles into effect are capable of improvement. Before undertaking the revision of "Printers' Cost," the Committee issued a circular and questionaire to members of the trade, and have received helpful replies. The replies showed a practically unanimous expression of approval in favour of a costing congress. With this cost congress being convened, the Committee are unanimously and cordially in agreement. They desire to express a most emphatic opinion that a proper costing system is absolutely essential for all houses, whatever the size, and that the present deplorable price cutting and variation in estimates is almost entirely due to the absence of costing systems. Until some fairly uniform method of costing becomes general, it is, in their opinion, hopeless to expect an increase in prices which should follow the constant increase in cost of production, and which alone can give an adequate return to the master printer.

The Committee are of opinion that the best means of obtaining the adoption of a uniform costing system is for some approved expert to be appointed, who will be prepared to advise on existing systems, and instal an approved system for a charge which would bear some relation to the size of the works. Before recommending such a course they suggest that the Committee be empowered "To prepare and issue a costing system with a view to its being laid before a Cost Congress specially convened for its consideration."

The Circular was as follows: The Committee of the Council of the Federation, which was recently appointed to consider the questions of costing and pricing, have under consideration the issue to the trade of further recommendations for the establishment of a uniform Costing System throughout the trade.

The Federation has already issued a pamphlet called "Printers' Costs," of which 9,000 copies have been placed in the hands of the trade, and the Committee are anxious to find out to what extent printers have adopted this or some similar system of costing, and what improvements or modifications of it can be suggested by those who have installed a costing system. It is worthy of note that Costing Congresses have been held in the United States and that, as a result, a system of costing has been drawn up on rather more elaborate lines, which has met with almost universal approval in America and is being largely adopted. The Committee believe that several firms in this country have installed the American system, or some modification of it, and it is hoped that these firms will give their experience of its working. The Committee will be gratefully obliged if you will favour them with any information you may possess, and with any views which you may hold on the question; and to facilitate this they ask you to be good enough to fill up the enclosed form and return it to the above address. The information and views placed on the form will be regarded as strictly confidential.

It is felt that much of the price-cutting which exists among printers to-day is due to an inadequate knowledge of the actual cost of production, and that some uniform method of costing on scientific lines would go far to remedy it.

MR. J. E. T. ALLEN (Manchester, past President), moved that the report, which was, he said, of the nature of an interim report, be adopted. Although they had given considerable time, and examined several systems, they had not by any means completed their work, and they asked for further powers, and especially for powers when they had got a scheme which would be acceptable to the majority of master printers to call a Cost Congress together to discuss the scheme with a view to its ultimate adoption. Their object was to try to evolve some system which would be so simple as to adapt itself to the smallest printer, and capable of such expansion that it should be also adaptable for the largest printer, so that each might find the actual cost of jobs, small or large, and might be able to balance his departments to find out what were the actual expenses of these departments, and whether they were paying or not, either weekly or monthly, instead of waiting until the half-yearly balance. He knew they had a very difficult problem before them in elaborating such a system and getting its adoption, but they had been able to make use of different systems already in operation, the Westminster system and the American system, some of which were in use by printers in this country, but their chief difficulty would be to get whatever system they decided upon generally adopted, because they would all agree that, if they were to get the benefit of any costing system, such a system ought to be uniform (applause). They would have difficulty in persuading those houses who already had a costing scheme of their own to change to another system, and he thought they would have still more difficulty with those houses who at present had no scientific costing system of their own, but were content to add some percentage to the actual cost of labour and material, with the hope that that percentage would cover all their standing charges, and at the same time leave a small balance for profit at the end. He did not know why it was that printers were so much afraid of realising what their costs were. Office expenses, rent, light, depreciation and insurance, were as much items of cost as were labour and material. It was only by realising that these costs must be met that they would get away from the haphazard method of estimating that existed to-day (applause).

MR. R. A. AUSTEN-LEIGH, who seconded, said he had a mania for costs, and he was inclined to say that the absence of cost systems was the root of almost every evil in the printing trade. He thought a good many of their labour troubles were due to the absence of cost systems. They were accustomed to under estimate, with the result that they had little profit at the end, and when labour made its demands they were almost always obliged to say that they could not afford to concede those demands. He believed that would not always be the case if they had a proper cost system. What they had to do was to convert the master printers who had no system, and who were quite apathetic. He knew that sometimes the half or three-fourths of the jobs were done at a loss, and he suggested to such people, who were making something out of it without a cost system, that if they had a cost system they would be able to eliminate things that did not pay, and get more for others, and their profits would be very much increased. He gave some details with regard to the Cost Congress which had been held in America, and which had resulted in a standard costing system being formulated. No less than 3,000 requests for this were received directly it was published, and in Boston, Philadelphia and New York, eighty per cent. of the output had adopted it, and there was only one instance on record of its being dropped. The master printers there were delighted with the system. Returns were made to the central office once a month, averaged for each district, and the resulting rate sent out to each master printer. After costing came estimating. An estimating class was formed, which was well attended by principals, estimating clerks and others. It would be an excellent thing if they could do something on the same lines in this country. They should have a uniform system, and it should be as simple as possible. They should have an expert able to advise them how to put the system into operation. The system need not be a very expensive one, neither need people be frightened about the trouble of putting it in.

He was sure people would be surprised and pleased with the results, and concluded by quoting the advice given by one speaker to the American Congress, "Put in a cost system before the Receiver puts it in for you" (applause).

MR. A. C. ROBERTS, the auditor of the Federation, returned thanks for his election as auditor, and also referred to the proposed cost system. He said the cost system meant economy in working. They would detect any leakage that might be in their business, or in any part of their work. That led to re-organisation when they found anything wrong through the costing. Following re-organisation there came the profits, and, after all, the cost system was put in for the purpose of better profits being made. The expense of working a system economically was very small, and was nothing compared with the benefits they would get from it. Experience showed that a proper percentage of profit could always be made when the master knew his costs, and had the benefit of knowing these costs, and had the strength not to let his customer make his price for him, which, he was afraid, was sometimes done in the trade. The main point in costing was to ascertain the hourly rate all through the house. Once they knew the hourly rate they must realise that they must get that hourly rate before they could make any profit. That was the trouble. Whatever they paid a man, a larger amount had to be put on to his actual wage before they made any profit. He was not dealing with the selling price, but simply with the labour output. He pointed out that he had acted for a good many printers in London, and he had put costing systems in. There was only one really good system, it was called the American system and the Westminster system - there was hardly any difference, really - and he had doubled, and in some cases trebled, the profits, although the same turnover had been made, after they had come in with the cost system, these profits being made by economy of working and the detection of abuses. He wanted all to understand that there was a little trouble at first, but directly the system was in operation they did not notice it, and their profits were very much more in their favour (applause).

The report was adopted, and the meeting unanimously approved of the proposal for a Cost Congress.

(*Members' Circular*, June 1912, pp. 162-167).

*

THE GREAT QUESTION OF COSTING.
(*1912*)

WITH THE UNITED TYPOTHETAE AT CHICAGO
Contributed by Mr. R.A. Austen-Leigh

To the English visitor at Chicago, whose chief memory of his own meeting at Edinburgh, last June, was one of four pleasant days of junketing, into which the intrusion of one serious morning's work made scarcely any effect, the American meeting was a revelation. True, much sound work was, no doubt, done behind the scenes at Edinburgh by Council and Committees, but at Chicago, not only were committees sitting well into the night, but for three and a half days, that is for seven sessions, the convention kept doggedly to work, either discussing some burning topic, or listening to informing papers on the printing problems of the day. And all this with the thermometer hovering fitfully around 100 deg.! No wonder it was a shirt-sleeves convention, coats being discarded by presiding officers and audience alike. Not that the lighter side was in any way neglected. On the contrary, what with reception committees, steamboat committees, theatre committees, automobile and banquet committees, every one was sumptuously entertained, but, except in the case of the ladies, all these entertainments took place in the evening when the day's work was done.

But to begin at the beginning, the Congress mostly assembled on September 2nd, at the Sherman Hotel. Over 800 members attended, so that it was all the hotel (which boasts some 750 bedrooms) could do to accommodate the lot of us. Not that Chicago thinks anything of such a convention: last year over 300 conventions were held in the city-an average of one every working day. There was no programme provided for Monday, beyond an informal reception in the evening, so that, after introducing himself to the courteous secretary (Mr. Heath), the visitor was free to explore the city.

Serious work begins with the meeting at 9.30 on Tuesday morning. Mr. J. Stearns Cushing, of the well-known Norwood Press, near Boston, more familiarly known as Captain Cushing (for has he not been Captain of the celebrated Honorable Artillery Company of Massachusetts ?) is president for the year, and calls on a minister who is present to invoke a Divine blessing on the conference. This done, the same minister, in a different capacity, namely that of an eminent citizen of Chicago, welcomes us all to his native city. To many people, he says, Chicago may still suggest something a little crude, and he reminds us of the wealthy but self-made Chicagoan, who some years ago took a book to be bound.- "How will you have it bound," said the binder, "in Russia, or in Morocco?" "What's wrong with Chicago ?" said the millionaire. "Why not bind it here ?" But Chicago has gone ahead much since then, we are told.

Next, the President of the Chicago Typothetae welcomes the United Typothetae to the city, and then Mr. Cushing is free to deliver his address. In welcoming us he makes special mention of the presence of two ladies who appear as master (or mistress) printers. Stress is laid too on the fact that this meeting celebrates the twenty-fifth anniversary of the birth of the organization, and that the first meeting, like the present, was held in Chicago. On that occasion there were sixty-nine delegates present. The organization had had its ups and downs, but to-day has a membership of almost twenty-five times the original number, and has disbursed an income in the past year of £14,000. Reports follow from the Chairman of the Executive Committees, and from the Secretary, and then banners are presented-one to Philadelphia, for adding the largest number of members to its Typothetae in the past year, and another to Richmond, Virginia, for the largest percentage of growth. This ends the morning's bill of fare.

We meet again at 2 p.m. and plunge at once into papers. First comes an eloquent address from Mr. H. P. Porter of Boston, entitled "The Father of the Man," dealing with the crying need for well-trained apprentices, and this is followed by a stirring talk on "the results that a cost system should bring," by Mr. Ellick of Omaha. Mr. Ellick is particularly strong on the indirect gains from the cost system. Thus by learning therefrom that the average number of impressions from his presses per running hour was, in reality, only about 900, he has by close attention been able to increase this to nearly 1,200. This paper begets considerable discussion, after which follow addresses on organization work in various parts of the country.

In the evening we get cool by making a pleasant trip in a large steamer on Lake Michigan, with a grand view of the lights of Chicago.

Wednesday morning is devoted to reports of committees, with papers following in the afternoon on such subjects as "Sick and Accident Insurance" - not compulsory as with us - "High Class Printing in the small shop," etc.

Finally comes the election of Officers, Mr. Glossbrenner becomes the new President, and the retiring one is presented with a medal, thereby joining what he himself calls "the honorary order of Pall-bearers."

In the evening we are taken to the vast Auditorium Theatre to see an admirable performance of "The Garden of Allah."

Thursday morning sees the Cost Congress open; this has now become incorporated with the Typothetae, but holds a separate session: Mr. Morgan, the Chairman of the Cost Commission, makes his report, and then Mr. Stone, of Roanoke, Virginia, is elected chairman for the meeting.

Resolutions are presented by the Cost Commission through Mr. Oswald, of New York, and considerable discussion arises. One member suggests that the handling of paper should be made part of the hourly machine rate, but this is not carried. Another member objects to adding 10 per cent. to the cost of paper for handling on the ground that it costs no more to handle an expensive than a cheap paper - this also is over-ruled. Finally the resolutions are passed with but a few verbal amendments, and no particular change is made in the system previously adopted by American Cost Congresses.

Papers follow during Thursday afternoon and Friday morning on such subjects as machine composition, costs and efficiency in the various departments. It is interesting to learn that Mr. Hartman considers the hourly cost of monotype keyboard and caster to be in the region of 10s. an hour, whereas linotype composition is usually reckoned to cost not more than 7s. 6d. an hour. Owing to the difference in wages these figures must roughly be halved to bring them to English level.

Finally about 2 o'clock on Friday afternoon the Convention is brought to a close, and the members adjourn with the feeling that they have by no means wasted their time in coming to Chicago. To the English visitor the most inspiring recollection is of the vast improvement, alluded to on every side, that has come to the trade from the adoption of a cost system. . .

Contributed by an earnest advocate of a proper Costing System.

This all important question is in the hand of an experienced Committee, which has spent much time and trouble in preparing a report for the use of our trade generally. There can surely be no doubt in the minds of any of us as to the urgent necessity of some united action, which will not only prevent such widely different prices being quoted for the same article, but which will, and this is all important, ensure our quoting only such prices as may be expected to produce a reasonable margin of profit. The differences in prices which we all experience, and which generally result in the order going to the lowest, at a non-remunerative price, cannot represent actual differences in cost of production, but are generally accounted for by errors in quality or quantity. Whilst there will always be a percentage of errors, yet the using of a carefully prepared costing system would go far towards preventing their occurrence. Perhaps it is of still more importance, that the system to be used shall be the same throughout the trade. For whilst varying systems might each be correct, yet, for varying classes of work, they would not give similar results. It is therefore to be hoped that, with a view to securing the advantages of a uniform method of costing, houses will adopt the system recommended by the committee, or vary their existing methods, so as to accord with their recommendations. This will involve time and trouble, but the results that may be expected, will far outweigh and pay for any cost and time spent in introducing an up-to-date system - a system which will not only show what print has cost to produce, but will provide an accurate basis for estimating purposes. It is expected that an opportunity will be given to the trade of learning all particulars, either at a Costing Congress in London, or at the large printing centres throughout the country.

Applications to the Secretary of the Federation for the assistance of an expert to lecture and explain locally to the members of any Master Printers' Association, stating that they were willing to pay so much towards the expenses, would strengthen the hands of the Executive Council, and possibly enable them to make the necessary arrangements to carry out a series of lectures throughout the country.

(Members' Circular, September 1912, p. 245-251).

*

THE LAUNCH OF UNIFORM COSTING

Having devised a system of costing for adoption by the printing trade, the Committee of Costs and Charges set about organising the event which would launch the crusade which was to herald a new profit-earning dawn for master printers. It was proposed to hold the 'First British Cost Congress' in London in 1913. As is evident from the material in the early part of the following chapter, the Cost Congress was heavily publicised in trade journals in order to ensure a good attendance.

The number of printers who were carried on the 'wave of costing enthusiasm' to the doors of the Kingsway Hall on 18th and 19th February 1913 exceeded the expectations of the organisers. The 1,200 attendees were addressed by the luminaries of the costing movement who provided detailed expositions of the cost-finding system designed for British printerdom and explained the reasons why it was urgently needed. The Congress concluded with the passing of a resolution to the effect that the costing system "be approved and strongly recommended for universal adoption by the trade". It was also agreed to hold annual cost congresses in the future and to establish a permanent costing committee whose function would be to promulgate the costing system and "organise a campaign of instruction".

The First Cost Congress was subsequently described in the Members' Circular as "the greatest mass meeting of printers ever held anywhere". A historian of the BFMP wrote in 1950 that "The enthusiasm of those gatherings was greater than anything before or since in the master printers' world and it heartened those committee members who had worked so hard and so long in the cause of costing" (Sessions, 1950, p. 49). The heady atmosphere is apparent from the main speeches at the Congress which are reproduced here. In February 1913 the expectation that the Congress would mark an epoch in the trade appeared to be no pipedream. In the immediate aftermath of the event regional cost congresses were held in the major printing centres of Britain to maintain the momentum of the costing movement.

THE COMING COST CONGRESS
(*1912*)

Great interest is being taken in the Cost Congress, which will be held, in London, late in February. The Costing Committee and the London Sub-Committee are closely engaged in the preliminary arrangements, and will gladly avail themselves of any suggestions from members, which may render it more useful to the Trade at large. The date has been fixed with a view to suiting the convenience of all concerned, but the Committee will not hesitate to alter it, if other dates should prove more suitable.

COST FINDING.
By "A MODERN PRINTER," *Member of the Costing Committee.*

This is the subject that is, or ought to be, the most prominent in the minds of all employing printers, for it has been much talked and written about of late, and the committee, which has been considering the subject, has already issued its first recommendations, and a Conference is being arranged for a date in February, 1913.

But, before and beyond all this, employers must have become painfully conscious that something, and something drastic, is necessary, if our trade, as a whole, is not to pass from its present unremunerative state to something far worse.

Is there a trade, during busy times like the present, in which the profits are so small in comparison with the capital, intelligence and energy that is required in the printing trade? Of the many directions in which improvements might be made, such as general re-organization, the installation of the latest machinery, efficiency in the management and staff, and the ascertaining of the cost of production, the latter is far and away the most important, and will best repay investigation, even if much time and money is spent on it.

What is the main object for which we printers spend our time and energy? An outsider, after enquiry, might be excused for saying to keep our respective plants and staff going, to increase the former turn-over, to quote as low a price as possible, and so on. Such an insignificant thing as PROFIT cannot occur to some in our trade, judging by the way they set out to make it, unless they hope to cover what they lose in one department by what they *suppose* they make in another; but this method was never proved to be successful, even with the man who ran swings and roundabouts, and certainly will not do in our day for a complicated business such as printing.

It cannot be necessary to urge the absolute necessity for preventing the reckless manner in which estimates are given to-day. So let us see how it can be prevented. First of all, no attempt is being made to curtail fair competition, but to get it on to such lines as will ensure some, if but little, profit. For it must be remembered that as the lowest price rules in all markets, it is the cut, re-cut and again re-cut prices which are governing us to-day, and which, for want of a proper system, are daily becoming lower and lower, whilst the actual cost of production is seriously going up. There can be no question of average - which, to some extent, might rectify matters - for it is only the lowest prices that are taken into consideration, fair and high prices being at once eliminated.

Is it realized that there is enough printing required in the country, which, if done with a reasonable margin of profit, would at once put us in clover? Our anxiety should be to make a profit on each order we undertake, and not to strive for additional work which, frequently, we can only obtain by undercutting the price of those who were previously doing it. Those who have tried the course of insisting on some profit know, as a fact, that although their gross turn-over has decreased their profits have increased, this is the course you are invited to try.

Now as to the method by which this much-to-be-desired object is to be attained. First, you must have a system, common-sense arrangement, call it what you like, which will show, amongst other things, where the leakages and losses occur, what each operation costs, what each order and also what each department costs, and see to it the costs are inclusive; and that you get these costs back *plus* the materials you use, the numerous expenses you incur, and a profit. Such a system as this is now offered to you already prepared, after most careful investigation and consideration. A system adaptable to any size of business, which no house, no matter how small, should be without if it wants to live itself, and let others of the printing fraternity live also.

This system will take time to instal throughout the country, but an earnest start is to be made at the Congress, after which, arrangements are being made for lectures to be given at all important centres, and experts engaged to explain and instal the system in those offices which invite such assistance. Each association should make early arrangements, through the Federation officials, for a preliminary lecture on the subject. In the meantime, its members can do much to improve the present position by immediately putting up prices-as every other trade has already done-to cover the greatly increased costs we are already labouring under. The circulars and slips, notifying an increase, that are being issued under the responsibility of the parent Federation, and which will be sent on application, should be largely used in each district, so that the movement may be united and far reaching.

But the dawn of the profit-earning era, which we are so eagerly looking for, will only arrive when the whole of the members are determined to bring it about, and is it not worth while acting in a united manner to achieve such a desirable result? United we can do it, just as our confreres in America have successfully done it. There the printing trade was in as bad a state as on this side, but after realising they were, more or less, existing to damage each other without improving their individual positions, they combined - as we must - to put the whole trade on a paying basis, and actually achieved it. Where they have succeeded so can we.

The members of your Council and of the Costing Committee cannot do more than strongly recommend the adoption of a costing system, or possibly the amendment of an existing one, the appointment of local costing committees for local governance of the trade, and the acceptance of the detailed system which they have provided; the rest, and it is the all-important part, the members of the trade must do for themselves, and do it in a united manner if it is to succeed.

By W. HOWARD HAZELL, *Member of the Costing Committee.*

The Cost Finding Committee of the Master Printers' Federation has decided to hold a Congress in London, at the end of January, to discuss this important question. The system they recommend, which is the result of many meetings and much investigation, has already been set out in a previous number of this *Circular*. It is hoped that, as a result of the London Congress, this Cost Finding system will be largely adopted, and similar discussions will be held in all parts of the country.

Owing to the rapid growth in the cost of production, our charges must be increased, or else a larger number of printers than usual will wend their unhappy way into the Bankruptcy Court. A bold increase of a fixed percentage on all our charges, the plan adopted by the type founders and many other manufacturers from whom we buy, would be the most satisfactory course, but this seems to be impracticable. The increased charges, therefore, must be made in detail, and a sound cost finding system is essential to find out those charges which are too low, and to convince the printer that he is making a loss on these jobs. The work of probably every printer can be divided into three classes: (1) jobs which show a good profit; (2) jobs which yield little or no profit; (3) jobs on which there is an actual loss. Unless there is a good cost finding system in use, the real costs on each job are not known, and the losses on some are hidden by the profits on others, and the printer is unable to find the weak spots in his business. All he knows is that at the end of the year he is only making a very moderate interest on the capital he has invested.

The Cost Finding Committee wished to find what was considered the average total cost (including all general expenses of every kind) of a compositor setting on time, and of certain common types and classes of printing machines. Identical questions were sent to a number of printers, and the answers were extraordinary, and proved the necessity for some uniform cost finding system. The cost of a compositor varied from 1s. to 2s. per hour, and there was almost equal variation in the cost per hour of the machines. It is not surprising that our customers find it pays them to send round for many estimates, when there is such doubt, or ignorance, amongst printers as to what it costs them to run their printing works.

The total cost of a printing works may be roughly divided into: first, the wages paid to what are generally known as "productive" workers, whose actual hours and wages cost can be charged against the work they produce; and, secondly, all the costs which cannot be definitely allocated to each job (*e.g.* foremen, clerks, travellers, rent, depreciation, power, interest, etc.), and it is these general expenses which are increasing rapidly, and which are so difficult to estimate. It is easy to find the value of materials supplied, and the actual number of hours the productive workers have spent on a job, but the difficulty is to know the exact amount to be added to cover all the great and growing general expenses. Processes, methods and machinery vary enormously in printing, and no uniform percentage can be added to wages cost. The percentage to add to wages cost to cover general expenses may vary from 50 per cent. to 300 per cent., and a good cost finding system is the only way to find out the correct amount. A printer who believes in putting on a fixed all round percentage on wages cost for machine work, will charge too high a price for his small and too low for his large machines. He gives a price based on estimate, and not on fact.

This is the essential point of the cost finding system. It shows easily and convincingly the real cost per hour of each different process. It shows it to each printer for his own works, not based on statistics or facts obtained in other works, and in a way he cannot disbelieve. It will show him the real cost on each job, and to which of the categories I mentioned they belong. If all the jobs show a good profit, let him be thankful and keep up his prices. The probability is, however, that he will find many "losers." If he finds jobs which cost him £25, and are charged at £20 or less, the master printer is not such a fool as to be content, but will mark down all "losers," and try to lessen his cost or to increase his charges. This will ultimately lead to greater uniformity in prices, less "shopping for estimates" by customers, and improved profits for printers.

There may be some printers who do not believe in cost finding on scientific principles, but I would urge them to discuss the question with a printer who has installed a good method, and ask him if he would like to go back to the old happy-go-lucky inaccurate ways. All those who have a good system realize its necessity, and are anxious that all printers should instal similar methods.

Whatever your opinions may be, come to the Congress. Listen to the explanations of the uniform system recommended by the Committee, discuss the details, speak of your experience of cost finding if you have a good method, and explain your difficulties if you have none. The result can only do good to the craft to which we belong.

WHY A COST CONGRESS IS NEEDED.
By R. A. AUSTEN LEIGH, *Treasurer of the Master Printers'*
Association, and Member of the Costing Committee.

All over the country printers seem to be waking up to a sense that something must be done to put their trade on a better footing. Trade booms come and go, but no boom seems to do the printer any particular or permanent good. Wages go up, the cost of materials goes up, but prices seem only to drop. Indeed the buyer of printed matter has become so accustomed to being able to get a lower price than he is at present paying by shopping round, that he has got into a frame of thinking that whatever happens; he is never to pay more than he does now; and if his printer dares to raise a price, all the customer has to do is to go round the corner to another printer, saying- "So-and-so wants so much for this job, if you can do it for less you can have it." And unfortunately there is so little co-operation amongst printers, and so little knowledge of costs, that in nine cases out of ten the neighbouring printer says to himself- "Well, if So-and-so can do it at that, it is worth my while to do it for five per cent. less."

Not, I am sure, that the customer wishes to pay less than a fair price for his printing, but naturally he wishes to be sure that it is a fair price, and so long as he gets, in open tender, prices differing by about 50 per cent., so long will he think the race of printers to be fools, and take advantage of any cut price that is offered him.

It all comes back to the two great *desiderata* in the printing trade - Co-operation and Costing. We want co-operation in order to get a uniform system of costing, and we want a proper knowledge of costs to help us to co-operate and not to cut into each other's prices.

The first step is to make up our minds not to sell a thing at less than it costs us, and not to try and make up for selling things below costs by selling a lot of them.

But in order to find out our costs, it is very desirable to have a standard or uniform system. Even then the costs will vary slightly in the same locality according to efficiency of management, etc., but such small variations might be very much accentuated, if no uniformity in the method of arriving at costs exists. That is one reason why we want a Cost Congress; namely, in order that it may adopt a uniform system.

Next what should be the chief requisites of a uniform costing system? Are they not that it should be as scientifically accurate as possible, as simple as possible, and as elastic as possible ?

It should be as scientifically accurate as possible, in order that it may be fair to printer and customer alike, so that when the time comes that we are able to assert that the cost of jobbing composition in London, say, is so much an hour, we need not, in the case of a dispute with any customer, hesitate to show the system by which such costs have been proved.

It should be as simple as possible, in order that it may be easily understood, and that it may be cheap to operate.

It should be elastic, so that it can be fitted to the smallest and largest businesses alike.

Now the system put forward by the Costing Committee and adopted by the Council claims to make good on all these three points.

It claims to be as scientifically accurate as is possible, and it assumes that the basis for such accuracy is:-

1. The division of a business into departments, varying in number according to the nature of the business.
2. The proper allocation to each department of all definite or direct expenses incurred by that department.
3. The allocation of all general or overhead expenses *pro rata* according to the total of each department's direct expenses.
4. The totalling of all chargeable or productive hours worked in each department.
5. The division of the expenses of each department (*i.e.*, direct expenses *plus* share of overhead) by the total *chargeable* hours worked in it, the quotient being the hourly cost.

Such practically is the system already adopted by the authoritative bodies of master printers both in the United States and in Canada, and although a system which suits America may not inevitably suit the United Kingdom, it is doubtful in this case whether a better could be devised.

The system, further, is simple, because little is needed except the proper allocation of expenses to departments and the totalling of chargeable hours; and it is elastic, because the principles are the same, whether a business has three departments or twenty.

Note, that it is the *chargeable* hour and not the *paid* hour; for so many hours are paid for but cannot be charged to the job (*e.g.*, clearing in the case-room, or waiting time in the machine-room) that it is of no use to find the cost of the paid hour, but it is the cost of the *chargeable* hour that we want.

Again, it is desirable that if the members of the Congress approve of the system, they should all go home and instal it. It is especially desirable that one or two members of all local Associations should attend the Congress, for we want them to go back and get their associations to take it up. If only Members will fill in the final form (Statement of Cost) results can then be compared, tabulated, averaged and fixed in each centre. For that is at present all we can expect to accomplish, namely, that printing costs should be fixed in each locality. We cannot, of course, suppose that the costs in London and Edinburgh are going to be exactly the same, any more than the costs of Edinburgh and East Anglia; but what we may arrive at is a regular cost in each printing centre.

And having arrived at a proper cost, the next step is to stick to it, and to have the courage, not only never to go below it, but to add a respectable profit to it. Let us see to it, that every job, even the biggest, bears its proper proportion of profit. Let us see to it that even the Government, that takes such an interest in fixing hours and conditions for us, at least pays a fair price, just the same as any private customer. Above all, let us cease from giving away to Paul what we earn from Peter.

I have no space here to dwell on the enormous advantages in efficiency and management that accrue from a cost system. It must be enough here to say that leaks are stopped, unnecessary hands dispensed with, excessive machinery reduced, or not renewed; in fact, that by such a system the finger of the management is kept perpetually on the pulse of the business.

And when we have reached co-operation and uniformity in costing, there are still further heights to climb. We want, above all, uniformity in trade conditions; we do not want one printer keeping printed stock or standing type free for a customer, and another charging rent. We want uniformity in charging for author's corrections. We do not want one printer to be throwing them in for practically nothing, and another to be charging *ad valorem*. We want, in fact, uniformity on such matters all over the country.

But it will be at least one great step in advance if we can get uniformity in costing systems, and a general installation of a uniform system.

Finally, if you know your cost and stick to it, you will find, not only a great increase in material prosperity, but in self-respect. You will gain the confidence and respect of the customer, who nowadays too often treats the printer with contempt, and you will be able to face any just demands of labour - and we must not forget that, as the prices of living and the standard of comfort increase, we are bound to be faced by such demands - if not cheerfully, at least sympathetically, and not with that hopeless and unsatisfactory cry of "we cannot afford it." I do not pretend to know what the attitude of the trade unions towards a Uniform Costing System may be, but of one thing I am certain: that if they have any sense and foresight, they should welcome it with open arms, and do all in their power to get it installed in any office where they have influence.

(*Members' Circular*, December 1912, pp. 331-339).

*

THE COST CONGRESS
(*1913*)

The holding of the first British Cost Congress marks a new epoch in the history of the printing craft.

During all the years in which the art has been practised no organised effort has been attempted, in our country, to bring master craftsmen together for the purpose of educating them to regard profit-earning as an exact science; and, unfortunately, so very few have troubled to seek a knowledge of costs that the whole business has, of late years, degenerated into mere huckstering, totally unworthy of what should rightly be regarded as an honourable calling.

In the United States and Canada, where labour demands have been rapidly becoming more and more insistent, and competition, until recently, increasingly keen and uncompromising, it was realized that if printing were to remain a profitable calling it would be necessary to take prompt measures to bring together master printers in order to discuss the best means of checking the cutting of prices to vanishing point. The outcome was a Cost Congress, addressed by some of the most enlightened printers of both countries, who advanced the theory that 90 per cent. of the craft were ignorant of what the work really cost them to produce. The subsequent discussions manifested the truth of the arguments adduced.

And the result? It is not too much to say that the costing system in vogue in America and Canada has revolutionized the methods of the craft in both countries. Prices have been raised to a reasonable level. Good profits are being earned, notwithstanding the granting of increased wages and shorter working hours to the men, and the printer is regarded with favour by the banks, whereas a few years ago financial assistance was sought in vain by anyone running a printing plant.

Surely we, in this country, have arrived at the breaking-point! Competition is becoming more and more unreasonable. The demands of our workpeople are becoming more and more onerous. The cost of raw material is increasing month by month; taxation is going up by leaps and bounds.

The remedy is now being offered. Instal a costing system, and learn your cost. The Cost Congress will make clear what a costing system is, and the Costing Committee and their experts will help you to instal one.

The Cost Congress will be attended by master printers from England, Scotland, Ireland and Wales. Small employers will be as cordially welcomed as large ones, and it will be open, also, to works managers, overseers, cost clerks, and book-keepers, indeed, to everybody in the trade, who is anxious for enlightened assistance in the work of making the business he is engaged in a profit-earning concern.

There is only one hope for the printing craft, and that is increased prices. There is only one way to secure increased prices, and that is for every printer to know his costs.

A special invitation will be sent to every member of the trade, and members are earnestly asked to interest their friends and neighbours in, and induce them to attend, the Congress. If any have not received invitations, ask for some to send them.

(Members' Circular, January 1913, pp. 16-17).

*

OPENING ADDRESS AT THE COST CONGRESS
(1913)

COLONEL WRIGHT BEMROSE, President of the Federation, who presided, after welcoming all present on behalf of the Council and Costing Committee, said he had received numerous letters of regret for non-attendance, and it was gratifying to notice that one and all were sympathetic with this movement, and some of them expressed the advantages they had already derived from working a cost method of their own, whilst many added that they would be very pleased to contribute towards the expenses of inaugurating this system.

Colonel Bemrose then proceeded: Gentlemen, I desire to explain that the committee who are responsible for the arrangement of this Congress, decided that I, as President of the Federation of Master Printers for the United Kingdom, should be in the chair to-day. (Hear, hear.). Had I known at that time the responsibility I now have to face I should have had to give more careful consideration to their proposal.

However, I very fully appreciate the honour, and it will give me the greatest possible pleasure if I can do any to forward the subjects which are of such vital importance to our trade. I feel, therefore, that now that we have met together, you will give such careful consideration to the suggestions about to be made to you that we shall be able to arrive at a weighty and useful decision, a decision which will be useful to us, I am sure, in the conduct of our business in the future.

Your attendance here shows that considerable interest is now being taken in the question of Costing and in the necessity of some action being taken to increase the profit-earning capacity of our trade. The difficulties we realise to-day are nothing compared with those we shall have to face if we do not speedily amend our present methods of doing business.

I have an idea that we should not congratulate ourselves too freely at this stage, for should we not have taken up this ill important question of Cost-finding long ago, before our difficulties became so acute? Let us rather make up for lost time by being thoroughly in earnest now we have met to consider the present unsatisfactory state of our trade. I do, however, congratulate those houses-and they are steadily increasing in number-who

HAVE PUT IN A COSTING SYSTEM,

and more particularly those of them, who, after realising the benefits to be obtained, have so persistently placed their views before our trade, and have at considerable personal inconvenience, worked the whole side question out for the guidance of the trade generally.

Before we can put our trade on a better footing we must have closer and better business relations with one another-(loud hear, hear's)- mutual respect and confidence within the trade, and what the Americans call "horse sense" in our dealings. I look to a friendly exchange of ideas during this Congress. Let us at the commencement fully realise that no costing system -no matter how good- will work itself. It must be carefully

ADAPTED TO THE PARTICULAR BUSINESS,

and introduced without haste, so that the staff may understand its working, and the employees realise that it is not against them, but for their interest as well as for that of the house. Whilst the introduction of a system will soon be of immense advantage in the working of any business, yet we shall not realise our larger anticipations of a better state of things generally unless and until costing systems are adopted very extensively throughout the trade-let me emphasise this. It will be

better for the trade if a large number of houses
ADOPT SOME SYSTEMS,
although they slightly differ in detail, than if only a few houses adopt
a uniform system, although I consider that uniformity is of great
importance to us. We must have combination and unanimity of action
if we are to succeed.

This movement is largely an educational one. We can all
learn something, for scientific costing, apart from estimating- they are
two separate things- is largely an accountant's business, and has
consequently often been neglected by printers. Let us look out for the
best suggestions that may be offered, and assimilate all there is to be
got out of them.

The knowledge of what others are doing will enable us to act
together, and this is all important in any trade.

Need anything further be said to convince us of the necessity
of accepting the Committee's suggestion, as amended at this Congress?
We have all suffered from the present unreasonable practice of careless
estimating, and the cutting of prices, a practice which aims at seeing
how little we can do a given work for, instead of ascertaining its proper
value-a practice which

CONTINUALLY RESULTS IN REDUCING FORMER PRICES,
and must go on doing so, irrespective of whether they were formerly
remunerative or not; a practice which takes no account of the continued
increase in costs of production; a practice which is quite worthless,
haphazard, and misleading; a practice which does not even give
satisfaction to the customer.

We must remember it is the lowest prices which rule the
market. We all know of cases where undercutting has only resulted in
spoiling the price for the firm who have previously done the work,
because they frequently retain it, but at a lower price. How, under such
circumstances, can good service be given to the customer? We are
often under the delusion that it is our customers who reduce prices.
May I suggest

IT IS WE OURSELVES
who hammer each other's prices down-(hear, hear)-and the customer
merely takes the favourable opportunity of getting his work at lower
prices. There need be no fear, on the part of our customers, that the
introduction of a system for the better working of our businesses will
materially raise prices. There will still be far too much competition to
allow of this. But no one should take exception to the competition

being on fair and reasonable lines. Some increase in price must inevitably occur, as in all other trades. We printers are not magicians, we cannot evade the universal increase which has occurred in the cost of production in all products, nor will our customers expect it. What we must attempt is

TO GIVE A BETTER SERVICE

to our customers, and this can best be obtained by the disclosure of defects in management, and more accurate allocation of cost to the individual order. May I, for a moment, also look at this question from

THE EMPLOYEE'S POINT OF VIEW ?

He should realise that it is well to be employed in a trade which is on a healthy and sound basis. He himself pleads that commodities have increased in price; he also knows that wages have constantly increased, and that hours have been reduced. These increases in cost affect printing, as well as other manufactured goods, and it is therefore obvious that all reasonable steps must be taken to put the trade in a better condition to meet these additional costs, and that the prevention of anything injurious to the trade is to his benefit as well as to his employer's.

WE ASK OUR EMPLOYEES' CO-OPERATION,

if only for the simple reason that a successful trade is in a better position to give a more sympathetic response to the applications of those it employs. What may we expect from the introduction of such a system as is suggested ? First of all, to obtain the necessary data on which to scientifically manage our respective businesses, with a view to making each order, each department, and the business as a whole, produce a profit. To enable the management

TO SELECT SUCH CLASSES OF WORK

as the particular business finds it can profitably produce, and to reject such work as a costing system shows it cannot, for one reason or another, produce at existing low market rates; in short, to prevent orders being booked at a loss. To prevent continued losses, arising from under-estimating, for if the system does not prevent errors of judgement (*i.e.*, estimating of time), it does at least show them as they occur, and thus prevents their repetition. And the gradual extermination of the present suicidal tendency to scramble for orders at any price, and so secure ourselves against ridiculous prices. To-day we are much too prone to accept orders at prices *we are told* some other firm will do the work at- an altogether unbusinesslike action. Reliable costs would give us more reliance in ourselves, and so we should be less swayed by the figures of others. Our keenness to obtain orders should not be allowed to over-ride our judgment of the fair value of the work. The anxiety to obtain an order should be in proportion to the profit it is likely to produce. (Hear, hear.)

The increase of our profits is the one thing we should work for, instead of too frequently considering the turnover as more important. All businesses are valuable in proportion to the profit they earn, irrespective of the amount of work they do. Better management means profits. It by no means follows that a business doing a large turnover and involving a large and continuous capital expenditure for plant and buildings will make more profit than a smaller one. We must look elsewhere for the profit-producing factor. The Costing Committee do not suggest that you should do something they themselves have not found satisfactory. They are offering you advice

BASED ON THE RESULTS

of their own systems. The suggested scheme may therefore be taken as the best general one that can be produced, although it may require some modifications to suit varying businesses. If it is open to improvement, the discussion we are about to enter upon will show in what particulars it can be improved. The Committee confidently believe that good results may be anticipated by those who put the scheme into operation. This course will take some thought and expenditure, which will be nothing in comparison with the beneficial results that may be expected. There are members of the trade, our fellow craftsmen, who could tell you, and will tell you I hope to-day, that by the introduction of a system they have actually increased their profits, although their turnover has gone down in amount. What individual members can achieve is absolutely nothing compared with

WHAT THE TRADE CAN DO

if it will only act in a united and systematic manner on this question of costing. United action in this matter would undoubtedly draw us together in other ways, and such a result in itself would be invaluable to us all. May I suggest that we forget the days when it was admitted, and actually accepted by some, that the composing room could not be expected to pay ? When work was taken to "fill up" irrespective of price, and when it was a practice with some to offer to take work at so much per cent. less than another house's price, leaving the customer to fix such price. If we cannot give up these old and expensive habits we must not expect any good results from our present attempt to put our trade on a better basis. I hope you do not anticipate that we shall arrive at a final conclusion by attending one Cost Congress, but we expect to arrive at useful decisions. We may not end all our troubles now, but we may - and will, if you so determine-very much improve matters relating to our trade: but this is for you to determine. (Loud applause.)

(*Report of the First British Cost Congress*, 1913, pp. 3-4).

*

THE SYSTEM EXPLAINED
(*First day of the Cost Congress, 1913*)

Mr. Howard Hazell: I have been asked to explain the system we have evolved as the result of our deliberations as the most suitable for printers generally. We considered the American cost system and we have had the advantage of Mr. Roberts' assistance, and have considered various principles in operation in various houses of members of the committee. From all these systems we have picked that which we thought was best and we do think the system we are recommending to-day is suitable for adoption over the whole trade. It can readily be adapted for businesses of two or twenty departments, or for printing works where there are ten employees or where there are a thousand. It is adaptable and flexible, and it must be adapted by you to suit your own businesses. In speaking this afternoon I shall try to imagine, if such a thing is possible, that there are printers who have never even considered any cost system at all. Some experts may consider my remarks very elementary, but if so I offer that explanation at the outset. At the same time, let me tell you that I heard the other day, of a man who knew the cost of composition, he knew that he paid 8.1/2d. an hour, and he added 1/2d. for general expenses, and that made 9d., and his composition cost him 9d. (Loud laughter.) That is the form of ignorance we wish to abolish.

THE CENTRAL POINT OF THIS SYSTEM

is that you should find for each man or each machine an hourly rate, or price per hour, which will cover the cost of wages and of all those general expenses which are such a bugbear in the printing business.

You will notice particularly in this form (No. 1), that there are only three departments, whereas in the forms circulated in the hall there are many departments. The totals in the final column are the same as on your form. For simplicity's sake we have only taken a few. It is simpler to deal with these only on this diagram. (The forms used by

Mr. Hazell were sent with others to all the printers in the country and were distributed in the hall).

The first step is to make out this statement of expenses. The figures filled here have been adapted from a business, and the first thing is to divide the total expenses from the profit and loss account into three great departments. In this particular business £16,000 per annum is paid in wages, the expenditure on material is £12,000, and there are expenses, including engineers, liftmen and paper warehouse wages, totalling £12,220, so the total expenditure is £40,220. The first thing to do is to divide the business into as many departments as necessary. In this particular business I think there are nine departments, of which here we only see three - composing, machining and binding - the principal manufacturing departments. There must also be a department for paper and materials. Having decided the departments, you then should fill in in the first column the various items of general expenses from your profit and loss account. This form only deals with general expenses totalling £12,220. Here there is only one composing room column because in this particular business there is only hand-composition. If you have hand-composition and lino.-composition or other mechanical composition, it will probably be best to divide the composing room into two or three departments, to find the expense of each process.

It is wise to take the expenses of the previous year, or if that year should be an abnormal one, take the average of the previous two or three years, so that the expenses may be average expenses. If during the year under consideration you find that your expenses have been increased you should increase your expense figures for the current year by the known increase. If your rent last year was £200 and for some reason it has been raised to £300, it is foolish to base on £200 that which you know has been increased to £300.

I will go through allocation of these expenses; to begin with, rent, rates and taxes. Here they amount to about £1,800, and you divide them in accordance with the floor space of each department. (At this point Mr. Hazell's long jointed pointer became disconnected and in pausing to attend to the "Repair Department" he quite took the fancy of the audience by the nonchalant manner in which he hoped, as he said, that his argument would be stronger than his fishing rod.) You will find (he continued) the square feet of the floor space in each department, and the square feet of the floor space in your offices. You should divide the total square feet into your rent and rates, and find your square foot cost. Next take Light, Heat and Water. It is probably best to divide these also according to the floor space: or if you like to be very accurate, according to the cubic capacity and number of gas jets or electric lights. Probably the floor space is the best basis of division.

I come to Power. In this composing room there were no linotypes. The bulk of Power was in the machine room, and a little in the binding; and in this diagram form you will see how the whole department is dealt with. In any case all the expenses in the first column must be divided into the various departments. Next comes Depreciation, a thing the printer often overlooks. We recommend that 10 per cent. written off the diminishing value of plant is a reasonable rate to take. Possibly 10 per cent. may seem rather high. A machine which cost £100, in twenty years, if depreciated at 10 per cent. off the diminishing value will be worth about £10 to £11. Some types of machines will wear out in less than twenty years; some will last longer than twenty years, but the Committee is of opinion that on an average 10 per cent. depreciation is fair and reasonable.

You will find the different values of your plant from your inventory, if you have an inventory. If not, take it from your balance sheet, and divide it over the departments, and write 10 per cent. from the value of the entry in each department. Fire Insurance will be figured according to the value of the buildings and contents, and on the basis of floor space, and with regard to stock in these departments. You will take your figure from previous stocktakings, and debit to the department the value of the insurance of the stock. Workmen's Compensation you will divide over departments where it occurs, whether on wages or by whatever method you choose. As to Repairs or renewals,

NO SENSIBLE PRINTER

would charge his Capital account with any repairs on plant. It is assumed that the expenses of repairs are charged to working expenses. If you have no details of previous years' repairs you must make entry in the column of estimates. Direct Departmental Sundries like proving paper machine oil and similar items that you cannot charge to a customer on every job must be treated as an expenditure made by each department, and here again, if you have no details of last year, divide according to an estimate.

Interest at 5 per cent. on capital-£30,000 in the given case-the Committee consider as an expense. (Applause.) I am afraid too many printers overlook this interest on their capital. If your business increases you will probably have to have a bank loan or issue further shares or debentures. It is unwise to look for nett profit until you have charged that definite expense of 5 per cent. on the capital employed. Let us suppose-you withdraw capital from your business and put it into some Stock Exchange investment, you would be able. probably, to get 5 per cent. on your money, and all your days could be spent in golfing or other enjoyments. (Laughter.) This interest item finishes the general expenses that you can divide over departments. Salaries and Commissions come next. Don't forget to write a good salary for yourself. (Laughter.) Your manager won't work for you for nothing. If you control your own business, write your own salary in this column, and your manager's and clerks' salaries, and your travellers' commissions, telephones, postages, and miscellaneous expenses come here. It is difficult during the course of the year to divide certain expenses such as those of wages for engineers, liftmen and paper warehousemen. So, at the end of the year you take their wages and put them in departments where they have mostly worked. Paper warehousemen naturally go into the paper column, etc. The total for the year is divided by 52 and that gives you the weekly standard departmental expense, and gives it to you for each department. In this way we have dealt with £6,810 of the total expense of £12,220, leaving £5,410 still to be dealt with. The point is, How are we going to get back this balance, £5,410, of overhead expenses? We recover it by adding a percentage on to the value of chargeable materials (*i.e.*, 10 per cent., or as the printer may ascertain for his own business), and by a percentage (to be ascertained by each printer) on the departmental cost, *i.e.*, wages, plus departmental expenses.

It would be well for me to define what I mean by "chargeable" and "non-chargeable." Chargeable labour is the time a compositor spends in composing, imposing, and correcting a job; the time of the machineminders in making ready and running, and the time of folders, sewers, packers, &c. Chargeable material is the ink, paper, strawboards, cloth, etc., used on a job. Non-chargeable labour is storekeepers, overseers, readers, time spent in distribution, etc. Non-chargeable material is proof paper, lye, roller composition, string, glue, and other similar items.

There are two ways of dealing with the overhead expenses on material. The first is for each printer to find out the percentage he should add. Already in the Paper or Materials column are placed the expenses of handling. He will go through the remaining overhead expenses, and see what proportion should be transferred into the materials column. He will transfer the travellers' commission on material, a share of clerks' and other salaries, to cover the cost of buying, invoicing and dealing with the materials. He will deal similarly with discounts, bad debts and spoilage. The total of the materials column will show the percentage he must add to the value of chargeable materials, viz., £12,000, to recover the share of overhead expenses that must be recovered on material.

The second method is to adopt the minimum handling charge of 10 per cent. recommended by the Committee. This percentage is only a minimum, and if the printer adopts the first method, he will probably find that the overhead expenses on material are more than 10 per cent. Ten per cent on the £12,000 worth of chargeable material in the case before you equals £1,200. You have already placed £571 of expenses in the materials column, so that the remainder of the £1,200, viz., £629, must be deducted from the overhead expenses. This leaves a little over £4,700 of overhead expenses which must be recovered by adding a percentage to the labour side of the business, viz., departmental costs.

You have in the total direct departmental expenses column £6,239; *ie.,* we are recovering a large part of the general expenses by way of distributing £6,239 among the departments. We are further recovering a smaller part of the general expenses by way of 10 per cent. (recommended as a minimum) upon the cost of the materials. After recovering all this £1,200 you still have to deal with £4,781 That £4,781 will be dealt with on the next form (Estimate of Expenses). This next form is made out once a year, at the beginning of each financial year.

In the bottom form (Form 2) you will notice that you see, total wages, £16,000. You have put into the departments £6,239 of expenses. You have to recover £4,781 of overhead expenses as a percentage on departmental cost. Find out what percentage you have to add to departmental costs to give you £4,781. In this case 21 1/2 per cent. added to Departmental Cost gives you £4,781.

Now see if that is correct. To test it you take the wage, £16,000; you put down £6,239 into columns as Direct Departmental Expenses, and that gives you £22,239, and 21 1/2 per cent. on this total of £22,239 is £4,781, and that with the materials £12,000 and the 10 per cent. on this materials, £1,200 will make up £40,220, which is the total expense of the business.

This shows that this system does interlock with the accounts of your firm, and that no expenses are overlooked. (Loud applause.)

Now to deal with the Statement of Costs of Production. Each week this form 3 is prepared, and first I take, in order, the amount of wages paid in each department. The wages book must be departmentalised, showing different totals of wages paid in each department.

If the foreman is paid monthly, you must add a correct proportion to each week's wages. You then add a weekly assessment of standing departmental expenses to cover rent, rates, wages, and the rest that I was just talking about. That amount you have found on Form 1, in the case of the composing room, is £35.

To the total of wages and departmental expenses you add the percentage to recover overhead expenses, which, in the case of this business, was 21 1/2 per cent. The result is the total cost for the week for running each department, which shows all the expenses of every kind, with the exception of the chargeable materials, which are dealt with separately.

The form is full proof, and cannot be doubted by the most sceptical printer.

Up to the present I have been dealing with the debit side of the business, and though Mr. Roberts will deal with the credit side more fully, I think it will be well briefly to speak of it. You now have the cost of running the departments for a week, but what is the value of production in each department for that week? Each chargeable worker, such as compositor, or machine minder, should keep a daily docket, on some simple form showing the hours worked on each job. These hours should be multiplied at the end of the week by the hourly cost. The total will give you the value of the work produced in the department, and the comparison of the value of production with departmental costs will show

IF THE HOURLY RATES ARE CORRECT.

If, for instance, you are costing your compositor at an absurd rate, at say 1s., you will find that the value of production is very much less than the cost in this department. If you are costing at 1s., and the total value is £100, and the departmental cost is £150, it clearly shows you it is costing, you 50 per cent. more than you have reckoned. If the cost of production is £200, and the value of production £100, it clearly shows you that your compositors are costing you 2s. per hour. In the same way compare various departments to see if you are correct.

In this costing system no week would absolutely balance the cost and value production. In this week (pointing to diagram) the Composing room shows a deficit of four guineas in Form 3, while in the machine room there is a surplus of over £4, and in the binders' room a deficit of £2 9s. In this particular business the surplus was a little more than the deficit. The hourly rates which it has put on the cost dockets were correct. One great advantage of this system is that it is a very convenient

BAROMETER OF THE STATE OF THE WORKS.

If you know that the cost for the week is £250, and you are only producing £200 this week, you know how much you are below your average; and if you are producing more you know you are extremely busy, and can then gauge the number of people you should employ.

Now as to the definition which the Committee have given of the words "compositors' hours." Some printers mean by an "hour of composition," the hours that comps., proof-readers, and pullers spend on a job. The Committee think it much better to have a uniform comp.'s hour, meaning simply the time the comp. takes to compose and impose and correct and send the job to machine; and to ignore the times of the readers, proof-pullers, overseers, storekeepers, and others, which will be covered by the compositor's hourly rate.

As to the time on distribution, that is particularly important in jobbing business, because when you take the time spent on composition it is impossible to find the time that will be spent on distribution, since the distribution may not be done for weeks ahead. Too often printers overlook distribution.

Having different ways of making up the compositor hour has this disadvantage, that one printer may be quoting at 1s. 6d. to include correction, etc., and another quoting at 1s. 3d. The 1s. 3d. man may get the job, and yet the customer might pay more. It seems best to have a uniform System compositor hour; which refers only to the time taken in setting and imposing.

Again, in the machine room the hourly rate of each machine should be its value, *i.e.*, interest on the capital value, the share of power and general expenses and the wages of the room, but not to include the amount of ink supplied. You may have machines worth 3s., which are running on a common job with ink worth 9d. per lb., and using very little ink, and the next job may be with an ink at 5s., and taking a great deal more ink than the job with 9d. ink. The risk is that the printer will overlook the value of ink supplied, and it is better to deal with the value of machines per hour, and not to include the value of the ink.

You will see that there is a Composing Room Analysis, which shows the history of the composing room week by week, as the weeks pass by. You will see if you are making a surplus or a deficit in the composing room. It is not wise to alter the hourly rates, till a long period has passed. Wait some months, till you have had a slack and a busy season, but you can see by that Composing Room Analysis whether the rates are correct or not. You should have similar analysis forms for each department.

There is another form, Form No. 5, which gives the Summary from the beginning of the financial year, to the date when the Form is made out. The only other Form that I wish now to refer to is the Job Cost Sheet, which will be fully dealt with by Mr. Roberts. On this job cost sheet you will assemble all the hours spent on this job. You will price them out at the hourly rates, found by the Cost Form, and you will be convinced by this that this is the correct cost of the job.

On to that job you will then add as much profit as the customer will permit-(laughter)-and your ingenuity will enable you to obtain.

Now to summarise. We have first the Statement of Expenses. On this are all general expenses from the Profit and Loss Account.

You make it out once a year at the beginning of the year and then only repeat the figures which you find at the bottom. Form No.2 - an Estimate of Expenses made out once a year at the same time, in order to find out what percentage you must add in order to recover your overhead expenses. Form No 3 - Statement of Cost. You make out weekly the wages paid each week; add the known share of departmental expenses, and add a percentage to cover the overhead expenses. Weekly also you get the value of work produced in each department. So you gradually find out whether you are correct. The Job Cost Sheet will give you the cost of the job at hourly rates. That is the labour side of the business. To this you must add chargeable material: ink, paper, etc., with its percentage to cover cost of handling this material.

Possibly some of you will think that this sounds as if you need an army of clerks. Those who have installed it find that this is not the case. It takes a good deal of attention to start it, but after that a junior clerk or a lady clerk can easily manage the matter. I don't wish to give you definite figures. I will tell on what my experience has been - and I think I shall be supported by other printers - that the cost of the clerical assistance is infinitesimal compared with the enormous advantage of knowing accurately what your costs are in general, and what are the costs of each job. (Loud applause.)

There are other advantages of this system. It does not interfere with your private ledger. It is not necessary to have an analysis day book, and an analysis journal. All you must do, is make out your statement of expenses. You can do this and the second form; or your auditor can probably do it for you, and if you wish to begin slowly you can start one department only-say a composing department. If you find it satisfactory, you can spread it to other departments.

It must be maintained. It is no good going on for a few months, and finding out a few costs, and then thinking you have done with it. One printer remarked: "I put in a good Cost System two years ago and now I know my costs, so I need not keep it up." Your costs two years ago are not your costs to-day. (Applause.) There was no Insurance Act two years ago; we were working longer hours two years ago. We were paying lower prices for material two years ago.

Week by week and year by year this system shows you clearly and conclusively what your costs are. It does not give you costs in anyone else's business, which you generally think is worse managed than your own. It gives you the cost of your business when worked out so that no sceptic can disbelieve it.

We are all in business for a reasonable profit. The first step towards this is to know your cost, and when you know this it is up to you to add a reasonable profit, and by salesmanship and ability to obtain good prices and increase your profit. (Applause.) I am afraid too many printers are buying composition at 1s. 9d. and thinking they are doing a clever thing by selling 1s. 4d.; and then they are surprised at the end of the year if the result is a bad one. They would not buy paper and sell it below cost. We believe this system will enable you to find your real cost in all your departments as easily as you find the cost of reams of paper. We do not tell you to raise all your prices to all your customers. We do not tell you what your hourly costs are: but we do ask you to put this system into your own works; we do ask you to find out   WHAT YOUR OWN COSTS ARE.

We say when that is done there will be less shopping for estimates, and less cutting of prices, and the printer's occupation will be a more pleasing one than it is at the present time. (Loud and continued applause.)

(Report of the First British Cost Congress, 1913, pp. 4-7).

*

ADDRESS BY THE FEDERATION'S AUDITOR
(*First Day of the Cost Congress, 1913*)

Mr. A. C. Roberts, Auditor to the Federation, and the originator of the Westminster System of Costing, was the next speaker. He was called upon to give explanations of some matters not dealt with by Mr. Hazell. He said: I am deeply sensible of the honour of being allowed to explain some part of the system which the Committee have decided to recommend. Mr. Hazell has given you a splendid exposition of the system, more especially with regard to the first and second part. The system is really divided into three parts. The first, any printer can manage without any assistance - the dividing up of his business into departments. Regarding the second, if he does not care about figures he can call in that necessary evil, he accountant-I am one-(laughter)- help him, and with the aid of the accountant he can complete Form 1 and Form 2, and he can get down to Form 3, where he sees the total departmental costs, and when he is there he can't get away from it? (No.) That which is recorded on Form 3 is what your department will cost you, whatever you do.

The matter I am going to talk about is the collection of chargeable hours. The first step is to settle what is chargeable and what is non-chargeable. The non-chargeable hour as a rule is waiting time. In the composing department distribution is non-chargeable. I have never met a compositor who wrote "waiting time." (Laughter.) He is generally dissing. (Laughter.) Every hour possible should be made chargeable. That is one of the first things to consider. In the Form in front of you you have a business divided into eleven department's. Whatever your business, you must divide it up. The Committee have decided only to recommend three departments to be considered to-day. Other departments can follow perhaps, at some later stage. These three departments consist of composing, machinery and warehouse or binding. The Costing Committee have decided to give you an easy system, one which won't bother you; they have decided to give you as few forms as possible, at first, to avoid any confusion in the minds of those printers who have never thought out the costing problem. As business doctors we are giving medicine in small doses. You can take one department at a time if you wish. Now then, to collect the chargeable hours. Every workman must write a daily docket; anyone who does any work that can possibly be charged to a job. It is recommended that a daily docket be used as preferable to a weekly docket, as being easier, very much easier. First of all, the compositor writes his name; he then writes the number of the job. He gives the customer's name; and then he writes the number of hours he is actually spending on composition, under the heading of ordinary time. If he is working overtime, he writes hours in the overtime column. If it is machine setting, he writes the number of thousands of ens. This docket every day is handed to the overseer of the department. The overseer will look over it and if he is satisfied, he should pass it down to the costing clerk. I will talk to you about the costing clerk; but let me say that I call this a Central Costing System, to distinguish it from the Departmental Costing System. I daresay many here let their overseers have a clerk to assist them in collecting the hours worked by their workpeople. It is very much cheaper to have a central costing clerk. I reckon a costing clerk and someone else could easily keep the costs of one business doing from £20,000 to £25,000 a year. I will tell you what the costing clerk has to do with the daily dockets. The Committee have not thought it advisable to give a diagram of a work ticket or order sheet. Everyone in this room uses a work ticket of some sort, that follows the work. The costing clerk makes out for every job a cost docket. He keeps the cost dockets in a loose leaf binder in front of him, and these cost dockets represent the work in progress. Every chargeable hour that goes through on a daily docket is charged up. This (pointing) is the composing room portion of the job cost docket. If your house has, say, eleven departments, you must arrange for a job cost docket to cover all these departments. If you cannot get them all on one form, get them on two forms. By charging every possible hour to the job cost dockets you will find all the hours spent in composing on each job. I should think pretty well everybody does something of that sort. Otherwise you would not know the number of hours spent on a job. But the mere collection of hours on a job will not give you any cost. We don't want to criticise any costing system now in operation, but I remember what someone said about beer, that there was no bad beer, only some beers are better than others. There are no wholly bad systems, but you may as well have the best.

Mr. Roberts proceeded to deal with compositors, and said: All compositors should write chargeable hours. The overseer is not included unless he can write chargeable hours. The total of the number of hours worked per day added up will show how many hours are worked in a week. You write overtime hours the same way as ordinary time. Hours spent on distribution are not chargeable in these columns (pointing to a diagram). (It might be parenthesised that it is impossible to convey in a report the way in which Mr. Roberts elucidated the subject when large diagrams were available, to which he could keep pointing with his long rod.) The hours spent on distribution are not charged to individual jobs, but distribution is part of the hourly cost. It is advisable to keep a record of the hours spent on distribution, so as to see what percentage distribution or non-chargeable hours bear to the chargeable hours. The total chargeable hours at the end of a week as shown by Form 9, converted into cash at your existing hourly rates, should be transferred to Form 3, Composing Room Column, and you will then find out whether you have been using in your business, you are proceeding satisfactorily or not. The Committee are not going to give you hourly rates at this Congress. They want you to work with

THE RATES YOU ARE NOW USING,

and to find if they are correct or not. You will get a deficit or a surplus by this plan. This system is one of double entry; you debit the customer with the hours taken on a job, and you credit the department. You cannot find whether your rate is right at the end of a week or a month. You can only find the sound hourly rate at the end of twelve months, when you have taken into account both the slack and the busy seasons. You can, however, make certain corrections as you go along. Paul Walmsley, who is a practical printer, has been working with me on Costs for many years. We have put this system into a large number of houses with most satisfactory results. When introducing a Costing System we frequently have to put in rates based on experience until we obtain figures which are reliable. We, however, always check all hourly rates in the way I have described.

I will now take the Letterpress Machines. I take it that all your machines are numbered, and that each machine works at a certain hourly rate. The machines' daily dockets are treated in the same way by the costing clerk as the composing room dockets. Your department is credited with the value of the productive hours for which Form 10 is used, and the customer is debited per job cost docket. From Form 10 you will see the number of productive hours run by each machine, which you convert into cash at your existing rates, thus obtaining a figure which is carried to Form 3 in the column headed "Machine Room." You will then compare the value of the output with the Total Departmental Cost. Mr. Cooke, of Leeds, will later on explain what is known as the unit system for ascertaining machine rates, so I won't say anything about that at present.

Now, In regard to the warehouse or binding department-you have a daily docket in front of you, on Form 8. This docket has been ruled to serve for both time and pieceworkers. It is called a daily or weekly docket. I think you will find it more convenient to use a daily docket for the higher rated labour, such as vellum binders, and more convenient to use a weekly docket for the cheaper labour. These dockets will again be dealt with in practically the same way as the compositors' and machine room dockets, with the exception that in the case of time workers hours are posted to the job cost docket, and in the case of piece workers cash is posted. The productive hours should be charged out at your existing rates, and the percentage you are now adding should be put on to the piece workers' wages. The total productive hours of each time worker and the cost value of piece work should be credited to the department on Form 11. You will then obtain the value of the production of the whole department, which is transferred to Form No. 3 and compared with the actual cost; a surplus or deficit is then disclosed.

Other departments are dealt with more or less in a similar way.

Mr. Roberts summed up and noted the point as to outdoor expenses being duly charged to jobs before being sent to the counting house, and also with reference to materials being charged up. The job cost docket is then completed and forwarded with the work ticket for charge to customer. In conclusion he said: I hope you will all decide to have the system installed. I hope the speakers will have convinced you that it is the best one to adopt. I do ask you to see it through regardless of the passive resistance of Know Alls. The Rule of Thumb system of costing has had a long innings. If you have no time to deal with a cost system, get someone to divide up your departments for you, to ascertain the chargeable hours and expenses and find their value. When you have found your value charge it in your estimates. If the rates come out higher than you have been charging, then charge the extra on to your customer. I don't think there are many in this room whose rates will come out lower than they have been charging. A proper system is a matter of vital importance, not only to yourselves, but to the trade as a whole. (Applause.)

(*Report of the First British Cost Congress*, 1913, pp. 9-10).

*

AN APPEAL FROM EDINBURGH
(*First Day of the Cost Congress, 1913*)

Mr. Blaikie, of Edinburgh, President of the Federation meeting lately held in that city, was called upon by the chairman. He said: I feel it a great privilege to meet so many brother printers and to have heard Mr. Hazell's masterly exposition of the Costing System recommended to you. I admired his address in answering puzzling inquiries, and feel if I am ever in trouble I shall go and put questions to him in private. I have listened, too, to Mr. Roberts' professional advice and his admirable statement, and now I presume I am asked to speak because it is thought someone must point out how necessary it is that printers should do something of the kind suggested. At the great annual meeting of our Federation at Edinburgh I pleaded for a strong union of master printers, in their own interests, and that of the trade generally, a union which should try to remove that distrust which has seemed to exist among ourselves, and which certainly exists

IN THE MINDS OF OUR WORKPEOPLE.

I pleaded for a strong union which should make our opinions respected. It has been extraordinary and encouraging to see in Scotland lately how such a strong union based on justice, has helped us to get over some difficulty we have had with our people, and to come out of that dispute able to convince our workpeople that our cause was just. For the first time in history we have come out having got what we went in to get. We found, Mr. Blaikie continued, that it was necessary to take Scotland as a unit of its own. We have not yet been able to take in the whole of Scotland, but Glasgow and Edinburgh are included in the unit already established, and in the trouble which arose (on a matter which was of great importance), Edinburgh and Glasgow joined hands. It was agreed that if Glasgow was attacked Edinburgh must join in, and that Glasgow should do the same if we were attacked. For the first time in the history of the trade we were obliged to issue notices. At the very last moment we were visited by the heads of the men's Printing and Kindred Trades Federation of the United Kingdom, and by their help we were able to stave off a lamentable strike and lock-out that had seemed imminent. I do not want to meet

MORE CAPABLE OR STRAIGHTFORWARD MEN

than the heads of the men's Federation. They discussed the matter with us, they saw where our cause was just, and they induced their constituents to settle the dispute. I do hope that that mutual distrust of ourselves and our work-people will be removed. It would be much better if we could trust each other and our workpeople, and state candidly just how matters stood. You may think this irrelevant, but the lesson I wish to draw by analogy is the power of strong union and mutual trust. If we in Scotland had not stood together in our troubles, we should have failed in our contention; and similarly, I have no doubt whatever that if we masters stand firmly together in a just cause and trust each other absolutely, we shall succeed in putting an end to that system of

SELF-SWEATING

in which we have indulged for twenty-five years. A knowledge of costing is the first step, for ignorance, cowardice, and distrust have caused that self-sweating. Ignorance of our actual expenses per unit, cowardice in exacting fair remuneration, and distrust of each other in estimating for contracts. More frequent meetings would help to remove distrust. If we could put our foot down unitedly as we have been lucky enough to do in the North in our late troubles, it would be a very good thing. With a costing system and mutual confidence I think we may begin a new era, and that self-sweating may be eliminated. I shall go back feeling very glad at having been asked to be present. (Applause.)

The Chairman: I have been asked the question: "What do the Committee recommend in cases where workmen will not make out time sheets?" That is a local question as to which there should not be much difficulty. One would think that it should be possible in any locality to make it clear to the employees that this is absolutely necessary in their own permanent interest.

(*Report of the First British Cost Congress*, 1913, pp. 10-11).

*

AMERICA'S PRINTING REVIVAL THROUGH BETTER COSTING
(*Second day of the Cost Congress, 1913*)

Mr. R. A. Austen-Leigh (Spottiswoode & Co.) was then called on. He said: I have been asked to speak on the subject of the results of a cost system,. especially as exemplified in America. And in order to do so, I suppose I am justified in assuming that everyone here has been convinced by the eloquent explanations of previous speakers as to the absolute necessity of having a costing system. And not only so, but also that everyone will take home the lesson, and convince the printers in their locality who are not here to-day, of the same necessity. We know that this is not quite an easy matter; there are various classes on whom we may have to spend a great deal of time and trouble. For instance, there is the type of printer who is too clever by half. *He* doesn't want any system: *he* knows that without any system his presses run faster than any of his neighbour's, that his comps. can produce more, and that he can buy his paper cheaper, etc., so that whatever price his neighbour puts in, he is always prepared to do it at 5 per cent. less - even without seeing his neighbour's price. Well, we have got to persuade that gentleman that not one of us, not even the youngest, is quite so clever as he thinks; that in the same town the cost of printing will come out about the same, except in so far as one office may be better organised than another, and that really the only way to arrive at that better organisation is by means of a cost system.

Secondly we have to convince the pessimist, the man who knows that he is always estimating at under cost, and that he is losing money, but who replies to all suggestions that he should put in a cost system, "Damn it, do you think I want to know how much I am losing on each job?" (Laughter.) Or who says "What's the good of trying to get a proper price ? The public won't give any more." Well, we have to turn that pessimist into an optimist by proving to him that printing is not a trade by which money must of necessity be lost, and that the public is not unwilling to pay a fair price, so long as it *is* a fair price that when once printers universally ask a proper price the public will have to pay it, or go without its printing - a very likely alternative. But as long as prices vary over 100 per cent. the public will no doubt continue to take the lowest.

Or again, there is the man who objects to the cost of keeping a cost system, or suggests that when once he has discovered his costs he should be allowed to drop the system again. Well, the reply to the first part is that the cost of keeping the system is only an insurance premium; *i.e.*, the premium you pay to assure yourself that you are running your business week in and week out at a profit, and not a loss, and that as such alone (leaving for the minute out of sight the resulting advantages it is the best investment you will ever make. (Applause.) And a reply to the second half is almost unnecessary because no one who has once installed a system and tasted the advantages of it, ever really wants to drop it.

Well, now, let us suppose we have converted these people. Then what next? For merely to have discovered your costs is obviously not everything. It is, indeed, a great step towards the uplift of the trade, I myself think the greatest. We all know the French proverb, *"Ce n'est que le premier pas qui coute,"* which I will translate quite inaccurately, "Costing is only the first step." It is, I say, a great, and perhaps the greatest step, but it is not everything. Even with a cost system you can still lose money if you have an inefficient organisation. It is an economic law that large profits can be permanently secured only by efficient organisation. How are we to employ the knowledge with which the costing system furnishes us? As to that, I think we are going to hear something interesting from Southport, a little later in this Congress. In the first place, we must examine our costs, and see if they are too high or not. It is well-known that everyone who finds out his costs finds them much higher than he thought. They may be higher than they should be. Two things may help him here; one is co-operation; by which I would suggest that in every centre a Costing Sub-committee should be formed from the Master Printers' Association, to which members should contribute their results, under numbers instead of names, if necessary. Before long it will not be difficult to arrive at some average cost which may be accepted

AS CORRECT IN THE LOCALITY,

and it will then be "up to" the printer whose costs exceed that average to use every means to reduce his own. But even before that, if a printer uses a cost system to the full, he will have learnt where his leaks are, and most of us have at least one big leak somewhere, if only we knew it. Thus your cost system will show you whether your percentage of unchargeable hours to the chargeable is too high; and if is much higher than 25 per cent., the chances are that some of your people are not exerting themselves to the utmost. It will show you how many hours your machines are actually idle in the week. And here let me say that one main result of the system in America was to prove

THE IMMENSE OVER-EQUIPMENT

of the trade. In fact, I heard someone in the States say that most offices were both overequipped and underorganised. Only last week I read in the Printing Trade News of America that in a large printing centre in the South conservative figures showed an over-equipment of 45 per cent. One word about this subject of over-equipment, for it is a serious evil. (Applause.) We all know the state of things that occurs periodically when we get a rush of work, the difficulty we have in getting it out, the amount of overwork we incur, until our foreman comes and says he really must have some more machines. Then we give way to the temptation of thinking the temporary rush is going to be permanent and order a couple more. What happens? The rush passes, and we are left with two idle machines. Now the sight of idle machines usually causes such a panic to a printer that he rushes round to find work for them, and the usual method of finding it is to cut prices-and the last state of that printer is worse than the first.

Therefore, unless you want your business to control you in the matter of prices, instead of your controlling it, think twice before you buy more machinery. Don't have machinery to meet your maximum demand, but have enough for your average demand, and see if

ONE OF YOUR NEIGHBOURS

won't usually be very glad to let you have the use of some of his machines in the time of your rush. Now, one part of your cost system should consist in showing you the exact number of hours in the year that your machines are idle, and I am afraid it often comes to a large number. Quite possibly you will find you have more machinery than you need. If so, don't go about cutting prices. Get rid of one or two of your oldest machines without replacing them. It will probably lead to immediate economies; not infrequently keeping an extra machine leads to keeping an extra minder; at any rate it may give you some valuable space. Get rid, I say: that is, either sell for their value, or better still break them up, for this prevents some competitor buying machinery cheap, and then putting in a low price because he has little interest or depreciation to charge. (Hear, hear and laughter.)

Again, your cost system should reveal to you the number of impressions you are getting out of your machines for every running hour. This usually brings another shock. The machine that you flatter yourself is giving you 1,000 impressions per hour (guaranteed by its maker, of course, to give you about 1,500) is probably not giving you more than 750. Possibly, by careful attention you might be able to increase this output. I heard a speaker at Chicago say that by giving this detail all the attention he could he had actually been able to raise his number of impressions per running hour from about 800 to 1,100. Think what this means in reducing your cost.

Here, then, are a few of the ways in which a cost system will show up the possible ways of reducing your costs *i.e.*, by diminishing the number of unchangeable hours in your case-room, by reducing the number of presses in your machine-room, or by preventing your buying new and unnecessary ones, and by increasing, your output per running hour.

In the next place, what are some of the advantages that a cost system will be found to give you? Let me tell you some of them.

It will give you the best grip of your business you have ever had, and further if will teach us all to be business men. Is not one of the chief evils of our trade that we have neglected the business end? We have not been business men - we have been good printers-excellent printers, I do not doubt, but we have been so much occupied with the detail of our business, have kept our nose so much to the grindstone, we have been so much in love with our trade as an art, that we have had to leave to the customer the duty of establishing the prices, and of settling the conditions of our work. (Laughter.) How else would it be possible for business men to consent to build large warehouses in which to stock their customers' printed matter free of all charge; to allow customers to send in their material, and charge nothing for handling and keeping it; to lock up capital in standing type for nothing, and to bear all the other burdens that we have allowed our customers to impose on us? Is that business? (No, and applause.)

But first of all the system is going to give us a grip of details. Before we have a system, we do not as a rule know what jobs are paying us, and what are not. We grope about in the dark. We do not know our costs, and consequently cannot be sure we are estimating sufficiently high to get a profit. But with a system you are going to know more about your business than you ever knew before. You will know what perhaps you didn't before, how much exactly more it costs to run a mono. per hour than a lino. - Perhaps you weren't aware before that every chargeable hour of linotype machine in London costs over 4s. an hour.

How needful it is to know more about your business you may judge from these seven prices which have last month been quoted for one public tender:- £140, £188, £201, £254, £304, £333-(laughter)-and £343. (Renewed laughter.) Two and a half-times the first man's offer. Is that business? (No, and applause.)

With your costing system you are going to discover whether working overtime, or double shift, is not really an economy instead of an expense. If, as I think, you will find in the case department your standing charges per chargeable hour are about 6d, it is obvious that while you are paying 4 1/2d. overtime

YOU ARE REALLY BRINGING DOWN YOUR COSTS.

Is not that something worth discovering? Does that not upset tradition? - Again, a properly kept cost system ensures your having a correct inventory of your plant-well worth while if you ever have the misfortune to be burnt out. Further it gives you a check on your wages. Let them be made out in the cost department or by the cost clerk. Each employee records his chargeable and non-chargeable hours; let your wage sheet record these, and see that you pay for no more than the total, after having checked the in-and-out time with that recorded by your Bundy clock. Such a system at least ensures that the workman records his whole time. See, therefore, that you get the value of what you pay for, *i.e.,* of your employees' time. I have heard it said that the dividends of certain cotton mills are all earned in the last fifteen minutes of the day's work.

Again, a cost system enables you to keep a proper check on materials. Keep your stock-ledgers in the cost office. Add the invoiced amounts of purchased stock, and deduct the amount of paper given out on jobs. See that the balance as shown by your books agrees with the amount revealed by your periodic stock-taking, or know the reason why. Again, a cost system renders it an easy matter to

TAKE OUT A MONTHLY PROFIT AND LOSS

account. Everyone, of course, knows the amount of their monthly sales, but without a cost system there is always an uncertainty about the amount of one's unfinished work. With a cost system there is no difficulty at all. The unfinished work is represented by the amount of hours on your current cost cards, or in your current cost books, which hours you turn into cash at the rates you discover to be your costs. The following is a short cut: Each month you add to last month's unfinished work the number of hours that have been added during the new month, and deduct the number that have been taken off the cost cards and charged to completed work tickets. The balance is your unfinished work.

Similarly from your stock ledgers you can find the amount of your stock in hand. And let me advise you to put away each night your cost cards and your materials ledgers in a fire-proof safe, so that if you do come back next morning and find your place burnt out, you will have no difficulty in proving to the insurance companies the amount of work you had in hand.

These, then, are a few of the advantages that a cost system is almost bound to bring. But you may naturally ask: Are there no disadvantages? And a very reasonable question is: Will the installation of a cost system mean that I shall get less work? Well, it is no good denying that if you follow up the discoveries of your cost to the logical conclusions, which are to see that every job, produces a profit, and that no new work is estimated for unless at a profit, it is no good denying that at first you will find your volume of work diminished. But a word of advice may here be useful. If a cost system proves that you have been doing a lot of regular work at a regular loss

DO NOT AT ONCE THROW IT OUT,

unless the loss is shown to be enormous. But use all your efforts either to find some more economical way of producing the work, or if that is impossible, see if your customer, who we may suppose is a customer of long standing from the fact that you have been doing the job for some time, will not be reasonable and allow you an enhanced price for it.

But as to new work, see at least if you cannot register resolve that in future except at a profit. The resolve has got to be taken sooner or later, if your condition is to improve, and therefore the sooner the better. Never mind if at first your volume does decrease a little; volume is not everything, and if the work you are getting bears a profit now, instead of a loss as previously, it possible that when you make up your accounts you will find that notwithstanding a smaller turnover, you can boast of a larger profit. And after all, it is profit that we are after.

And a word here with regard to our old friend, the fill-up job, or the filler, as Americans call it. What is to be our attitude towards it? Well, if you will take my advice, you will avoid it like the devil. (Laughter and applause.) No doubt it needs some courage, and I am not prepared to say that there are never circumstances which may not authorise your taking it. But there is this much to be said against taking it, that if you take one you are very apt to take another. Secondly, that if it gets known that printers take work at cheaper rates when they are slack, the cunning buyer makes it his object before placing work to find out what firms are slack just at that time. Again work which you may estimate below cost because you are slack, may perhaps not be put in hand actually till much later, when you will be very busy again.

Well, then, avoid the fill-up job, but if the temptation is too strong and you feel you must have a try for it, resolve, if you like, to forego your profit, but be sure not to go absolutely below cost.

Lastly, what have been the main experiences of the system in America? I believe I am right in saying that the costing has

REVOLUTIONISED THE WHOLE OUTLOOK

of the trade. A few years ago the state of the trade was just as desperate as it is over here to-day. Competition was of the most cut-throat description; every printer mistrusted his competitor if he did not hate him; customers got just what prices they liked; the master printer was despised by the whole community as essentially an unbusinesslike person. Till one day some of the most sensible and level-headed men in the trade came to the conclusion that only by some proper costing system could a start towards better things be effected. Three and a half years ago the first Cost Congress took effect. A committee was formed which formulated principles. These were adopted, and a Permanent Committee set up to draw up forms, etc. The movement caught on to a wonderful extent, till now Cost Congresses are taking place in almost every State. I was told last September that over 1,000 firms had put in the standard system. The great organisations viz., the Typothetae and Ben Franklin Clubs, send out experts who go to every centre and give lectures on how to instal the system. And to-day an entire new spirit may be said to reign in the trade. This has been effected partly by costing and partly by co-operation, but mainly by costing, for it is costing that has brought about co-operation, and that has shown how far superior co-operation may be to competition. Competition is good enough for a time, but when it has reached a certain point, then co-operation is better.

The American master printer who has had the intelligence to put in a costing system, is a different man from what he was before. He has set himself a new and a better standard; a cost system has shown him that he has a right to a salary just as much as any of his employees, to a salary that counts as an expense, and not only to a salary but to a respectable profit. That as head of a business, it is not for him to overwhelm himself in a mass of detail that he can afford to pay others to do for him. The saying "Never do yourself what you can afford to pay others to do for you," is not such a bad motto in printing as well as other business.

Nowadays the American master printer controls his business by the grip he gets on it, through his costing subordinates he has time to devote to salesmanship and to planning large campaigns. He has learnt, too, the benefit of taking regular holidays. For if a boss can't trust his business to get on for a month without him, you can be sure that there is something wrong about that organisation.

"I don't know how it is, but I don't seem able to leave my shop for half an hour without everything going to pieces," said a master to a friend calling on him; and the friend replied: "Well, if so, it seems to me you have one man in this place who

DOESN'T KNOW HIS JOB."

(Laughter.) And not only does the proprietor benefit by the increased respect his banker and everyone else pays him; but his foremen benefit from the system by the greater ease with which they can carry on their duties. And lastly, there is the workman, and I am glad to think that he stands to benefit quite as much as anyone. When once the printing trade has grasped the elementary proposition that the public has got to pay the cost of printing, together with a decent profit, surely half the terror of adding a shilling now and again to the workman's wages, as the standard of comfort increases, as we all wish it to do, will speedily disappear. Wages are being increased in the States, I believe, because of the costing system, quite as much as in spite of it.

If this be so, the whole trade, master, overseer, and workman, can assuredly combine to bless the movement. It would be a blot in the system if the master only tended to benefit, and not the workman. Let us, therefore, do our utmost to impress this fact on the workman's mind. It is true that the workman can do something, to hinder costing if he tries-he cannot stop it, but he can hinder it by objecting to fill in daily dockets; if he won't do it it must be done for him by the foreman who gives out and receives back the work. But let him once understand that he stands to benefit just as much as anyone else, and he will see how

IT IS TO HIS INTEREST

to help in the movement. For the printing trade has come to this position, that there is little more that can be wrung out of the profits for the benefit of the workmen, until the public is made to pay a fair price for the printed product.

Finally let us not forget that we owe a considerable debt to America: those who like myself have had the opportunity of studying their system at first hand, and of attending one of their Cost Congresses, are not likely to forget either their hospitality or their willingness to help the inquirer. Our system, as Mr. Hazell has explained it, is very similar to theirs, but it is not entirely the American plan. We have been able to study the American and other systems and select the best points. I am sorry that we are late in starting after the Americans. Let us see if we cannot overtake them, first by learning to cost properly, and secondly by using that knowledge to estimate correctly, and finally by developing a higher power of salesmanship. (Loud applause.)

    (Report of the First British Cost Congress, 1913, pp. 18-21).

*

WHAT A COST SYSTEM HAS DONE FOR ME
(Cecil B. Johnson (Ben Johnson's, York),
Second day of the Cost Congress, 1913)

With your permission I propose to read what I have to say, as I should probably miss out half the points if I did otherwise. My position reminds me somewhat of a converted sinner, who tells of his past misdeeds to the admiring crowd at a revival meeting. In days gone by, I believe it was the practice of the Spartans to make their slaves drunk in order to show the rising generation how foolish a man looked in that condition, and the Costing Committee have paid me the compliment of putting me up as the awful example of a man who has put in a Costing System.

Perhaps it would be as well to let you know why I have been considered a fit and proper person to talk on the subject. It is because we have had in actual operation for three years the Cost System under discussion.

Towards the end of the summer of 1909 my brother felt the need of a rest, and, after packing a suit case, he hied him to America, where he wandered for six weeks, living largely on the generosity of our brother printers across the Atlantic. (Laughter.)

He left home with the fixed intention of forgetting that he was a printer, but the hum of the press and the smell of the ink proved irresistible to him, with the result that he found himself being shown the sights, entertained to lunches, and generally treated like a king by the master printers of the various centres he visited.

In addition to the sights, he was shown, the Cost Systems used in the various offices that he inspected. The point that struck my brother very forcibly was, that he found essentially the same system in a majority of the offices that knew anything of the subject. The suit case developed into two, the second containing an ever-growing pile of cost forms, and my brother returned to England convinced of the advantages of the American system. He found me mildly interested in the idea, but not enthusiastic, as we had, what I naturally considered a satisfactory system-I had developed it myself. (Laughter.)

The working of the old system was as follows:- After fixing hour rates for every operation with due consideration and discussion, we charged everything out at these rates and added a percentage to cover overhead charges and profits. At the end of the year, our auditor drew up a statement of cost, showing a comparison of value of production as shown by our rates and the actual cost. From this, we learned the minimum percentage that we could put on to cover all the costs, and no one, with the exception of ourselves, knew exactly the point at which the cost ended and the profit began. This we thought an advantage, but we have realised since that if anything

IT WAS THE REVERSE,

because it gave the impression that we were trying to get a much larger margin of profit than we actually were, and our travellers had no great confidence in our quotations, as they did not know what they represent

To cut a long story short, I was persuaded to go thoroughly into the new system with a view to adopting it, and after nine months' work, we had all the forms drafted and were ready to start. This may seem a long time, but it is explained by the fact that the whole thing was new to us, and the forms brought from America covered only the letterpress and pamphlet departments, and so we had to extend the scheme to cover every department we had, and we did not dare to make an actual trial until we had every link in the chain complete.

The installation was simpler than we anticipated, as the suspicions of the men, and the anticipated antagonism of the foremen did not give additional complication; in fact, after working closely on it for a fortnight, both my brother and I left England for a holiday, and the Cost Office was allowed to take care of itself.

Here I may say that I know that all the forms were sent to Manchester for inspection by the leaders of the Typographical Union, almost before I saw them myself-(laughter)-while the foremen were distinctly opposed to what we were doing, but we explained that, as we were going through with it, it would be better not to condemn it until after a fair trial, the result of all which was that there was no friction whatever.

After six months, we got out a sheet based on the previous year's expenses and current wages and hours, which proved to us that certain of our rates were too low, while others might be reduced. In other words, we were losing money on some classes of work which we thought were paying, while others were carrying the loss. A little consideration convinced us that every other printer we knew anything about was

IN THE SAME HAPPY STATE,

and so we decided to reject all work, the price of which did not cover our new rates, and try to replace it with other work, which we knew other printers must be doing at paying prices without quite realising it. (Loud laughter.)

A colossal impertinence, you will say, but events have proved that we were correct in our opinion. At this point I should like to reply to a question which was asked from the back of the hall yesterday afternoon, the reply to which was probably not satisfactory to the questioner, as Mr. Hazell evidently did not hear exactly what was said.

The question was: "What provision does the system give for a workman to give an explanation of his non-chargeable time?" by which I understood the questioner to mean-what opportunity has a workman of justifying his non-chargeable time in cases where he feels it is excessive.

On Form 6, you will see a column headed "kind of work," and if a man has lost time in picking, let him put picking in that column. Let him write it in red ink and underline it if he feels so inclined, as it is just this information that we wish to have. If a master does not provide material for his men to work with, he must not blame them for losing time in looking for it !

And now to the advantages of the system, so far as we are concerned: -

The men like it because, while there is no loophole for cooking time, they realise that it is absolutely fair, and no attempt is being made to play one man off against another.

The foremen like it because it saves them trouble as the booking, that they used to do, is now done for them in the Cost Office, incidentally more accurately and at less cost, because clerks are not usually paid as much as foremen. (Laughter.) This gives them more time for oversight. Further, knowledge of the actual output per hour for each machine helps them enormously with the estimating.

The management like it because work is traced with a fraction of the trouble experienced previously. Materials are dealt with satisfactorily-we no longer have the same stock in two or three different places and we know that the time taken can no longer be "averaged" or guessed.

The travellers like it because they have confidence that our prices are correct, and they know that if others are 20 per cent. or 30 per cent. below us the work is not worth having.

The directors like it because the profits are larger-(hear, hear)-and the anxiety is less, while the fact that our turnover has increased steadily since the system was installed, proves that our customers like it. (Applause and laughter.)

The last sentence is the reply to those who fear that the system means loss of work - we certainly have lost a number of jobs on which we were losing money, but our turnover has actually increased since the introduction of the system.

In conclusion, I beg to say that Mr. Lake is responsible for the title of my paper, and in reply to him I should like to say that the Cost System has not "done for us" -(loud laughter)- nor does it appear likely that it will do so.

I have been asked what the working of the system costs us. It is approximately 1.42 per cent. on what I pay in material and labour. (The 1.42% includes accountancy, wage books, stock ledger, and monthly balance sheet work. Costing staff alone would cost a good deal less). The increase in my gross profits has been 25 per cent. (Applause.)

(*Report of the First British Cost Congress*, 1913, pp. 22-23).

*

AN EDUCATIONAL MOVEMENT
(*Second day of the Cost Congress, 1913*)

Mr. J. R. Riddell (Principal of St. Bride Foundation Printing School): I feel it to be an honour to be requested to speak at this large and enthusiastic gathering of master printers, met to discuss how best to place their businesses on a sound commercial basis. At first sight it may seem strange that as a Principal of a technical institution, I should be interested in the correct ascertainment of cost, but gentlemen, this is an educational movement, and as such, the technical instructor must be in close touch with it. If you will allow me, as it were, to drop my academic role, and speak as one who has made a study of cost systems, and who has a practical knowledge of the American standard uniform cost finding system, I should like first to say how much we all have enjoyed Mr. Hazell's exposition and answers to the various questions. These were dealt with with all the subtlety and suppleness of the practised politician -(laughter)- and as the individual who put the question yesterday referring to "picking," - that "picking" should not be called "a non-chargeable cost" but should be called bad management,- let me say that the answer I got was one which, whilst not a direct answer to my question was the admission which I had set myself out to obtain. I hardly know if this was due to Mr. Hazell's natural cleverness, or if he simply caught on. I angled for that declaration as to bad management. Such leakages are undoubtedly to a great extent bad management, and it was pleasing to note how rapidly the audience responded to Mr. Hazell's way of presenting the idea. It gives one great hopes for the future of the working conditions of the printing office. At present there is no doubt such items of bad management exist in a very large majority of our printing offices.

I can also heartily support Mr. Austen-Leigh in what he told you about the American system and what it has done for the American printer, but I should have liked him to tell you what I found whilst in the United States. It was this: that many American firms spent a great deal of money in introducing accountants into their workshops to evolve a costing system for them with the result that many of them found that owing to the technical nature of the work it was necessary for the printers to combine and work out their own salvation. This they have done, giving a lead to the printers of the universe.

Now, gentlemen, I said that cost finding was an educational movement, and it must begin with yourselves. The principals and heads of a firm must first be efficient before they can expect their workers to be efficient, and towards that end let me say that it would be a mistaken policy to place the working of a practical cost finding system into the hands of an inexperienced man, no matter how enthusiastic he might be. Remember that enthusiasm without experience will make any system unwieldy, and one that will ultimately defeat the purpose for which it was created. Allow me to refer to a point which few speakers have touched upon. It is advisable, when introducing a cost system, that the aims and objects be

EXPLAINED TO THE WORKERS.

There is nothing to be gained by forcing on to intelligent men a new system without their having a knowledge of what it is instituted for. It would be useless to say that there is any recording system popular with the average worker, but this unpopularity is often due to the method of introducing it. Again, the daily docket is only a small part of a cost system; still, it is very necessary that it should be accurately dealt with, and these records ought to be looked upon as a basis for the ultimate betterment of the workers, and this view should be pointed out to the men. Referring to the system placed before you, whilst personally I should have advocated a closer resemblance to the American system, where the men have the opportunity given to them

TO DETAIL THE MANY HINDRANCES

met with in the printing office, still I would like to emphasise the need for a uniform system, which can be made to suit the requirements of the various offices. Mr. Chairman, I think that the splendid enthusiasm which has shown itself within these walls will not evaporate as soon as this great gathering disperses. At this moment it speaks well for future co-operation and harmony, with a greater belief in the sincerity of the intentions of one's fellow printers. And when the master printer obtains a proper return for his labour, skill and investments, he may then have more time to devote to another vital point, *i.e.*, the practical and technical training of the apprentice.

(*Report of the First British Cost Congress*, 1913, pp. 32-33).

*

CLOSING SPEECHES
(*Second day of the Cost Congress, 1913*)

The Chairman proceeded: I see on the agenda that the Chairman will now "summarise the conclusions arrived at during the proceedings." (Loud laughter.) I see you appreciate the difficulty; the more so, that we have not arrived at any conclusions. (Renewed laughter.) But I think I shall be safe in saying that we do appreciate, and we are all in accordance with, the very excellent and explanatory addresses which we have had during this congress, and the very pertinent questions that have been asked. I feel we have gone very much further and more deeply into the details than we could have hoped. I hope when all this matter is re-printed that the subject will be taken up very vigorously at all your centres. A question has been asked-What is the Council going to do to inaugurate a movement in the various centres? The Council will act-providing you give the necessary funds, and that is a very important matter. I must refer to it. (Hear, hear.) We cannot carry this through as we ought to do, and as we should wish to do, unless we have financial support. That is really necessary. But in any case the London Council will arrange for expert lectures for the large and important associations in the kingdom. Many of the associations have already sent their names as desiring to have such lectures. They don't want to be left out in the cold. We must leave it to you who have been in this Congress to pursue the matter further, to attend the lectures when they are held, and to get others interested. I am sure you have been interested yourselves, and if beyond the inaugural lectures you want further lectures or instruction, you will, I hope, establish local costing committees, and arrangements can be made through our worthy secretary for expert help in the following up of the matter. Experts will go into the country and render their help for a fee of five guineas, and they are prepared to go to groups of employers or to individual employers. Or you can have a report upon your business, or on what you are doing, or on what it would be advisable to do-you can have this by private arrangement. We have to leave that to the chairmen and secretaries of local associations.

If I am to make any general remarks on the subject, I would like to impress that the unit, whether it be a time unit or a piece unit, must be the whole cost. It must include the whole expense of the department -everything but profit. Then I would say, make a very careful distinction between chargeable and non-chargeable hours. These are the vital points. What is "chargeable" is one thing, and what is "non-chargeable" is another. I think some of us in the past have failed to grasp what has been involved in this distinction. In regard to that 10 per cent. on materials, let me remind you it was suggested as a minimum, but does not include profit, and it should be added to the paper you handle, although it may belong to other people. On the matter of depreciation of 10 per cent. on diminishing values of plant, that has been proved to be a fair and reasonable allowance. In some places the Inland Revenue Authority agree to it; in some places they don't. It is up to you to show them that it is reasonable. The machine unit suggestion is most reasonable. It is a matter you all want to work out most carefully. A question arises as to an alteration brought into the case by the introduction of new machinery. New machinery costs more, and that fact alone is the ground for just consideration by the customer. Sometimes you get a machine that does more operations than your old machine; you get one that prints and folds, or one that prints and numbers. If you get one that does two new operations, then I suggest you vary your rule, and consider that that machine is worth an additional amount, so that nothing be given away.

Cost finding is finding the cost per hour or piece unit. Estimating is the application of that figure to the hours you think or know that a job is going to take. And here we may have large differences of opinion. A costing system, if properly worked, should be a very great help and doubtful estimating or guessing should be done away with. All these advantages will accrue to those who put in this system. If we are wrong in our figures of estimate total, and right in our rates, then we shall know that the error is one time will remedy.

A question was also raised as to whether a general uniform adoption of the system is what we want. It has been shown to you that good as other systems may be, they vary in results. A great point with us is that we want a uniform system to be in as many houses as possible in this country-as many as we can possibly induce to instal it.

I am pleased to tell you that the Committee will endeavour to prepare a simpler scheme for smaller printers. (Applause.) That will be published for the use of all concerned. Also Mr. Lake has undertaken to have lantern slides of the various diagrams shown; and they will be at your disposal.

I think we may congratulate ourselves on what we have accomplished at this, our first Congress, and I hope the fruits of our labours will be plentiful. We must remember that the tilling of the ground will be necessary and that the seeds or ideas can by further cultivation, be very much improved. This we have to leave to you very largely, as I have said; and I do hope you will take it up locally, and where you can go into matters more deeply than we could. I do hope at you will be able, after careful consultation, to formulate operations upon hour rate.

I must strongly advise, that when you have arrived at that stage, you all adhere to the rates that are agreed upon. You can even go further-some already have done so-you can fix prices for some of the smaller work which is always being cut up, and get a selling list to date. It has been done with very great advantage in certain centres. In our effort to pursue this matter we need not wait for the next congress, which I am already assuming we shall have next year (Applause.) I suggest we carry on our discussion through the Circular, where questions and answers can find place. It is an admirable publication. I find very interesting matter in it myself. I hope in future we may be able to keep in touch.

There are other things, the Chairman continued, we cannot touch on to-day. For instance, we want to consider what terms of payment we should ask for when estimating. We want to consider the credit that is given and ought not to be given. We want firmer terms for corrections; we want better defined conditions and terms for which type will be kept standing. There is the matter of stocking of goods, and the very vexed question of free sketches. (Applause.) I now call upon Mr. W. A. Waterlow.

Mr. W. A. Waterlow: I have to propose to you the following resolution: "That the Cost System discussed by this Congress, with any necessary amendments, be approved and strongly recommended for universal adoption by the trade." One or two points have struck me in listening to these proceedings. First of all, that the Committee came before this Congress with a cut and dried scheme. We thought that was the best way. We submitted what after consideration we concluded was the best system to recommend, and I consider that the criticisms have been very slight indeed. There was this morning a criticism and a suggestion that another system was better, but I did not hear that supported by any other speaker. You will understand that we fully discussed all the systems brought to our notice during the eighteen months we have been working.

It was said by Mr. Bisset that what was wanted was to know the results particular firms who had tried this system had achieved. The house I am connected with installed a system on practically identical lines five years ago. It has given the directors and management the very greatest satisfaction. We would not be without it for anything. (Applause.) You can have no idea what added interest it will give you in your business. I speak personally. I take very great interest in it. I did not believe it was possible that I could have taken so much interest. I have found it real enjoyment. I can also speak of its profit, of its enabling one as one goes along to gauge very well the profit one is making. It gives you a grasp of your business that nothing else can do. There is one point about it, and that is its value if you ever come into dispute with your customers. I have no fear whatever with regard to any customer with whom we might be in conflict. With this system you would in any legal proceedings be able at once to show what was cost, what was profit. If you have had experience before judges or official referees or in county courts, you will know that it is a difficult proposition for you when you have only bare costs on to which you have to put a percentage to cover all those expenses You may find you have to put on 100 per cent. or more on your labour, and it is very difficult to convince any judge that that is a necessary percentage. If you go into court with these details, and show every item thus worked out and what it costs you altogether, you have a very strong position. Only in the last few weeks our secretary (Mr. Lake) had to give evidence in court on a question of profit. There was no proper system at work in that case, and it was an impossible matter to decide where cost ended and profit began. Mr. Johnson said it took a long time to instal this system. That, however, was a case where he worked up a system and installed it. You have this cut and dried, and ready to put into your business. You will not find it a difficult matter to instal it. Mr. Johnson gave you his cost of working it. His figure is in excess of what seems to prevail in other offices, but even so it is a saving. You withdraw clerks from other departments, centralise your costs. I feel deep interest in it, and am absolutely convinced if we are able to pursue it to the utmost it can effect an improvement in our trade nothing else can effect.

Mr. Jas. Forman (Nottingham): The great enthusiastic and earnest meetings of yesterday and to-day are convincing proofs of two things. The first is that a very large number of printers are satisfied that there is something rotten in the state of printerdom, and, secondly, that a very large number are determined to try to remedy it. In my opinion the only remedy is a satisfactory system of costing, and not any attempt at equalisation of prices, irrespective of the costs each house is incurring in a particular job. I would confirm almost every word Mr. Waterlow has spoken. Our experience in installing an almost identical system for rather a shorter time than Mr. Waterlow stated has been exactly the same as his. The remedy for printing troubles lies, in my opinion, in good costing. We have got to replace wish bones by backbones, as Americans say. (Laughter.) "Universal adoption" mentioned in the resolution is very desirable. I would advise that those who have no system should try this one. The man who has a system which he has found unsatisfactory will also probably be induced to give it up and try this. There is likely to be more difficulty in inducing a man to change a system which may be acknowledged in itself to be good. At the same time it is really most necessary to have uniformity. The result of having two systems will be this, that there may be marked differences in the cost of a job with heavy material and light labour and the cost of a job with lower proportion of material and higher labour. The customer will get the lower price in both cases, and the printer will only get the leavings. There is this other point as to why it is necessary to be universal. There is no printer so small that he does not react upon the printers with whom he comes into contact. (Applause.) Competition goes right out, right up to the largest printer in the kingdom. Therefore, we are all connected one with another. Therefore, it is necessary that every printer, large and small, should have a good system. As the adoption recommended in this resolution becomes general we shall possibly then find that even competitors are very good fellows towards each other. A better feeling may take the place of the suspicion and distrust of the present moment. These are some things which we may well desire to see, and which are likely to follow the installation of a good uniform costing system.

The motion was carried unanimously amid applause.

Mr. W. A. Clowes, President of the London Association, moved: "That the Costing Committee be warmly thanked for their labours, and requested to immediately promulgate the system, with any necessary modifications and forms, and to organise a campaign of instruction in all important centres by lectures and demonstrations by experts, and that a permanent Costing Committee be appointed to carry on the work, to consist of not more than twelve members." I wish (he said) on behalf of the whole of the printing trade, to thank the Committee for the immense trouble they have taken in getting all these particulars together with a view to laying final conclusions before us to-day-(loud applause)-and, what is still more important, on the spending of so large an amount of valuable time so freely and so generously for the benefit of the whole printing trade, during the whole of the last eighteen months. I can only hope their reward will be great and we shall all reap the benefit of their untiring zeal. (Applause.)

Mr. Theo May (Lewisham): I have great pleasure in seconding. The best practical thanks will be our conveying to the printing trade the information that has been given to the Congress. (Applause.) I take it there are gentlemen here convicted by the facts, and yet who still are in doubt whether they dare put the system in operation. They fear that their competitor round the corner won't do it. We can divide printers into three classes-those who are convinced as to what their cost is, and so will charge accordingly, irrespective of what happens; those who are ready to adopt a costing system if they think they can do it without losing their business; and, thirdly, there are those-we must recognise it-who will let others increase their prices and will try to get their customers away because of that. We must educate and also organise and make it impossible for that third man to get the business from those who want to trade legitimately. When I left the hall last night someone showed me a letter sent to one of his customers. The writer said they had a large amount of plant lying idle for a great portion of the week, and they were prepared to put in very low cutting prices on that account. Can we expect (Mr. May continued) to get our rights as man or master if we are not prepared to join an association that is doing its utmost in our interest? The Costing Committee has devoted an enormous amount of time to this work. We have had the experience and the unpaid time of some of the brightest intellects in the printing trade. (Loud applause.) It depends now on us. If we are unconvinced and undetermined, we are not going to put this plan into operation, and we are not in that case going to reap the advantage. It is essential that every master should come into our association. If there should be some who still lag behind, there are ways of bringing them in. There is need for this adoption of uniform system. One firm of printers - a fair-sized house - found their profit had been going down by £100 a year. They talked the matter over with someone who knew. This gentleman put in a proper costing system for them. Immediately this firm realised that the biggest job they did, the job they thought was the mainstay of their business, was one they were losing an enormous amount upon. They notified the customer, who got another firm's tender, which was 50 per cent. more than they had been paying. They had to pay this elsewhere, and eventually, as the new firm were not satisfactory, gave it back to the printers at 100 per cent. above the price they had been paying. (Applause.) That firm were doing at this day that recovered and now profitable job. Mr. May cited another case where a firm were proposing to quote £90; a man who understood costing persuaded them to charge £130, and that price they got. (Laughter and applause.) We have been in Egyptian darkness, and can now enter the promised land of fair prices; it depends upon ourselves whether we first wander in the wilderness for forty years. If we do, our sons won't. You can make this great improvement, Mr. May concluded, before the year is out, and you will enormously benefit by the result. (Applause.)

The motion having been carried unanimously and very heartily, Mr J.E.T. Allen (Chairman of the Costing Committee) responded: I know I am only voicing the feelings of all who have attended this most enthusiastic and splendid congress, when I say that we shall return home to instal a costing system, and to inquire thoroughly into the working of it. I think that perhaps I am entitled to a little measure of gratitude personally as chairman of this Costing Committee I have had some responsibility in saying who were to take charge of some subjects, and I think you will agree that the choice we made in

committee was a very wise one. I have been Chairman of the Costing
Committee since it was formed about eighteen months ago; and
although I have given a good deal of time and attention to the working
of the Committee personally, I can assure you the burden of time and
work has fallen mainly upon the members representing London, who
have formed themselves into a sub-committee and have attended many
more meetings than we were enabled to attend. I desire to thank them
most heartily for all the work they have done. Our thanks are due to
Mr. Lake, our secretary, for the help he has given us, which cannot be
overestimated. He has been splendid in all his undertakings from the
very beginning until now.

I just want to add this: You have spoken nice words; you have
attended the congress in large numbers. We think, as a committee, we
are entitled to ask you to prove that you mean what you have said.
(Applause.) . . .

. . . Mr. R. H. H. Baird (Belfast) moved that the Cost Congress
be held annually at a time and place to be arranged. In adopting these
reforms (he said) we are following the lead of our friends in America.
There they have an annual congress and not only so, but every printer
who adopts a costing system seems to become an apostle for converting
other printers. Our firm in Belfast have used this cost system, and our
last financial year was the most successful we have ever had. In
addition, we have now taken to marking time on dockets by clocks. We
lost some old and valuable customers-we used to consider them valu-
able-but our turnover instead of being reduced has been increased. In
Belfast we suffer very seriously from one or two firms who will not
come into line with us. They seem to think that other printers have
some treacherous scheme in their minds. I do hope all will try, in their
respective towns, to get as many printers as possible into line.

Mr. Herman Lea (Messrs Wertheimer & Lea): I beg to second
this resolution. The wise man profits by experience; the fool does not.
We have all been here for the last two days because our trade experiences are not satisfactory. Are we going to profit or not? Those who
have put in costing systems all testify one thing. They testify that these
systems bring them in increased benefits.

The motion was carried unanimously.

Mr. Sidney Reid (Newcastle) moved thanks to the speakers and
those who had read papers. He said: I have been recalling the days
when the Cost Committee was formed which ultimately produced
"Profit for Printers" and "Printers' Costs." I think Mr. Howard Hazell

was chairman of that committee, and did a great part of the work in producing those publications. We owe him a vast debt of gratitude, as well as for his work in the last few months. The sequel of the various labour troubles was that subject of costing more or less died down for a time. Our past experience tells us that we must not let the subject of costing die down. We must press it until every firm recognises that it cannot carry on business successfully without a definite system of costing. If you think for one moment you will see that habitually reckoning 18s. as satisfactory where you ought to require a pound means a loss of £2,000 on a £20,000 turnover. We must keep the subject before us till we have got away from reckoning our costs at one penny less than they really are.

Mr. Geo. H. Harrison, President of the Leeds Association, seconded. We cannot express our sense of gratitude, he said, for the labours of those who have come here to-day. As a recognition of their enthusiasm, I may quote that I heard at lunch a suggestion that there should be a stained glass window, with figures of St. Bemrose, St. Hazell, and the rest. (Laughter.)

Mr. Thomlinson, of Glasgow, supported. The very lucid explanations we have heard will, he said, remain in our minds, and I hope will bear fruit, and that in England, Scotland, and Ireland we shall hear of efforts being made to bring these things before the printers of the country. The most important thing is to get the actual factors of cost. Getting that solves all the difficulty. If a master with his eyes open will sell below cost, let him; but most printers are sane men. In Glasgow we called in a professional accountant; we sent two hundred forms with questions to printers. Had they been properly answered? We could have prepared a most valuable guide for all Glasgow printers. Two hundred copies were sent out; only three were returned. Yes, things have improved since, and I say: Don't be downhearted. (Applause.)

The thanks were voted unanimously, amid loud applause and cries of "Hazell."

Mr. Howard Hazell received quite an ovation in rising to make acknowledgment. He said: On behalf of the other speakers and myself I wish to thank you for the appreciation you have shown of our efforts. Nothing can give us greater pleasure than that good may come to the trade to which we all belong. We have given days, and some have given weeks, in the last eighteen months to these labours. All the

reward we desire is that the craft to which we belong may be benefitted
as the result of our efforts. (Loud applause.)

Mr. Siddall proposed thanks to the Chairman, who had presided,
as he reminded them, at the first Costing Congress. I venture (said Mr.
Siddall) to think it will-be a Congress that will mark an epoch in the
trade. (Applause.) I trust we have struck at the root of the old feeling
of distrust of which Mr. Blaikie spoke. As to the much discussed
ignorance of cost, the remedy is in a costing system, and the best way
in which we can show our thanks to the Chairman of this Congress is
the practical way of putting in a costing system forthwith. (Applause.)

Mr. Hobbs seconded. The name Bemrose is a household word,
he said, and Colonel Wright Bemrose is a worthy scion of a worthy
family.

The vote having been accorded very heartily,

The Chairman: I am much indebted to the proposer and seconder,
and to you gentlemen. My duties have been light and pleasant because
you have been so fair. Further the large attendance and patient hearing
and the interest you have taken throughout the whole proceedings have
been very pleasing to observe. But I do hope you will drop that idea of
the stained window- (loud laughter) -and rest content with the portrait
given you in the papers circulated in the Congress - a portrait
introduced without my permission by the way. (Laughter and cheers.)

An enthusiast started "Jolly good fellows" honours as a last
acknowledgement to the workers of the Congress. It was well taken up,
and amid all this hearty display of appreciation and good will the great
gathering broke up.

(*Report of the First British Cost Congress*, 1913, pp. 35-38).

*

THE COSTING SYSTEM

The costing system which was approved at the First British Cost Congress was published, together with specimen forms, in April 1913 under the title of The Printers' Cost-Finding System. *The booklet was authored by W. Howard Hazell, who was later to become the first Vice-President and Second President of the Institute of Cost and Works Accountants. The second, slightly more expansive edition of the manual, which was published in July 1913, is reproduced here.*

In October 1913, the Organizing Secretary of the BFMP Costing Committee reported that 8,000 copies of The Printers' Cost-Finding System *had been printed and sent "to practically every printer in the British Isles". He asserted that the purpose of so wide a distribution was not only to instruct employers in the technicalities of costing: "In my experience the booklet serves more to arouse or revive interest in the costing movement, than as a guide to those who have not heard the system explained, and sending these out, makes it all the more necessary to arrange meetings wherever possible" (Members' Circular, October 1913, p. 350).*

The costing techniques and concepts expounded in The Printers' Cost-Finding System *were not particularly novel but were subject to comment in the trade and professional journals. Most critical was* The Accountant. *The extracts presented at the end of this chapter show that concern was expressed about the lack of integration between cost-finding procedures and financial accounting systems. The Accountant also considered that the system prescribed for printers was "a half-measure" which was more reminiscent of estimating than 'real' costing. It also questioned the capacity of the average master printer to comprehend and implement the system.*

In general, the BFMP welcomed the criticism of the cost-finding system due to its potential for encouraging more widespread interest in the costing cause. However, the adverse (and uninformed) character of the commentary in The Accountant *resulted in a stream of correspondence to its editor from the leading actors in the costing movement.*

THE PRINTERS' COST-FINDING SYSTEM
(as approved at the British Cost Congress, 1913, with Specimen Forms)

PREFACE

THE PRINTERS' COST-FINDING SYSTEM is the result of over twelve months' work of the Committee of the Federation of Master Printers, which was appointed in October 1911 to consider the question of *"Increased cost of production, and the consequent necessity for Increasing Charges."* The Committee considered that if printers had a scientific system of Cost Finding, and based their charges on the actual ascertained costs of the different processes, the difficulty would be largely overcome. With this in view, the Westminster (systemized and named by Mr. A. C. Roberts) and various other methods in use in this country were considered, and Mr. R. A. Austen-Leigh visited America to investigate further the system that has been introduced in the United States with such beneficial results.

From the information thus obtained the Committee has evolved this system, which contains the best points of all the methods that were examined, and which was unanimously approved at the first British Cost Congress, held in London in February 1913.

Many enquiries were made after the Congress for a pamphlet which would fully explain the principles of the system. This explanation, has been written in simple language in the hope that a printer who has no cost-finding methods may be able fully to understand it, and introduce this system in his Works. Many points and questions that have been raised at recent Congresses held in various parts of the country are included in the explanations.

The Costing Committee desire to thank Mr. W. Howard Hazell for compiling this book in which the Costing System is explained.

All the members of the Committee have read the proofs and made many suggestions; and in addition, Mr. R. A. Austen-Leigh has written the Introduction. Mr. A. C. Roberts has furnished the original figures and much information from which the forms have been filled up, and Mr. Harry Cooke has prepared the Machine Unit form.

Any printer who desires further information or assistance should communicate with his local Association, or with the Secretary of the Federation.

INTRODUCTORY REMARKS

Before attempting to explain the Cost System that the Committee have prepared, and which has been approved by the first British Cost Congress, it may be advisable to answer certain questions that may be asked. In the first place-

WHAT IS COST ?

Cost may be defined as the sum of all the expenses, direct and indirect, incurred in the production of a given article. It is the exact point between profit and loss. Yet how many printers know the exact amount they make, or lose, on each job they carry out ?

WHAT IS MEANT BY A COST SYSTEM ?

A cost system implies a systematic method of discovering cost, as opposed to guessing. It means a certain amount of routine, for no proper system can be evolved which does not necessitate some clerical work and the keeping of various forms.

WHY A COST SYSTEM AT ALL ?

Because in these modern scientific days no manufacture can be carried oil successfully for any length of time without an accurate knowledge of the cost of the article manufactured. Printing, which is a manufacture as well as an art, is no exception to this rule, and the unfortunate condition of the trade as a whole may be considered as almost entirely due to ignorance on the subject of cost.

WHAT ARE THE, MAIN ESSENTIALS OF A PROPER COST SYSTEM?

They are that it should be (1) simple, (2) accurate, and (3) elastic. It should be simple, because simplicity implies ease in operation as well as economy in clerical labour. It should be as accurate as possible, for if its principles are not scientific, it will not command the assent of the trade, and therefore it will fail to secure that universal adoption which is so desirable. It should be elastic, in order that it can be adapted to large and small shops alike.

PREVIOUS COST SYSTEMS

Two alone need be mentioned: (1) The system by which an identical percentage, sufficient to cover all general expenses and based on the previous year's Profit and Loss Account, was added to wages and material alike. This is unscientific, because far too large a percentage has to be put upon material, and any printer using such a system would be quite out of the market when quoting against a competitor, employing a more scientific system, on an order containing much material and little labour. It is true that the same printer might successfully secure the jobs in which there was much labour and little material; but the tendency would be for him to secure this class of job alone, and he would probably, therefore, be much out of pocket at the end of the year.

(2) The system by which, material being practically taken at cost price, a percentage was placed upon labour sufficient to bear all other expenses. This, again, was unscientific, because it put all the expenses on to labour and none on to material. The result would be that the printer would secure those jobs in which there was a large amount of material, and lose those in which labour was the principal expense, and thus he would be unable to secure enough work to cover his general expenses.

The system now advocated is suitable for adoption by any printing office, whether large or small, and whether it has many or few departments. The various systems in use in this country and in America have been carefully considered by the Committee, and the experience thus gained has been of great value in preparing a system which should be easily adaptable to the many difficult and complex problems of a modern printing plant.

EXPLANATION OF THE SYSTEM

The Costing Committee of the Federation of Master Printers originally issued a Report and Recommendations, with a specimen set of twelve forms. It was suggested that if the forms were simplified and reduced in number, and issued with a pamphlet fully explaining their use, the system would be more readily adopted by printers.

The following explanations apply particularly to the forms for small businesses attached to this pamphlet, Nos. 1 to 12, and to the forms 13 and 14 explaining the machine units, which apply to large and small businesses; the principal change is the reduction of the number of, and the amount of information on, the forms. It would be well for a printer with a large business to obtain a full set of the forms as originally issued, so that he may know the additional information that he will find useful.

It is probable that when a printer installs this system, he will ultimately find it desirable to have the further information which can be obtained by the forms originally issued for larger businesses. The forms issued with this pamphlet have been found to be satisfactory by printers who have adopted them, but they must be modified and varied by every printer to meet the peculiarities of his business.

The essential points of the system are as follows:-

1. The cost per hour (if the work is done on time) or per piece unit (if the work is done on piece) must be found for each process in the business.

2. The hour or piece-unit cost should cover all the expenses of wages and general expenses (eg. idle time, rent, rates, taxes, selling expenses, and interest on capital, etc.).

3. A percentage sufficient to cover the cost of warehousing, handling, and selling should be added to the nett cost of material.

4. The individual cost of each job on the above basis should be ascertained, to which should be added a reasonable amount for nett profit.

5. The above hour costs should be automatically checked at recurring intervals, so that the printer may know that the hourly rates on which he is working are covering the costs of production.

STATEMENT OF EXPENSES (1)

In introducing a cost-finding system into any business, it is necessary to divide the business into departments. The number of departments will vary with each business. In a small business three manufacturing departments of composing, machine, and binding, and a materials department, would be enough, but to these must be added foundry, litho, ruling, etc., etc., as necessary. Each printer must decide for himself the best divisions of his business.

If the composing room includes monotypes or linotypes as well as hand composition, the composing room must be divided into two, as the cost of one thousand ens set on machine is very different to the cost of an hour's composition by hand.

If you have a large business, you will probably want more than one materials department. The cost of warehousing, handling, and insuring a large paper stock is heavier than the same cost for ink, cloth, etc., though the selling expenses will be the same.

These forms have been filled up with the approximate figures of a small business with a turnover of about £3,000 per annum and a capital of £2,000. These figures are inserted, not as any indication of what the hourly rates should be, but merely to assist in the explanation of the forms. In the business taken as an example there were three departments-composing (on time), machine, binding (on time), and materials.

Each printer must fill in these forms with the figures from his own business to find his own costs.

The Profit and Loss Account, form 15, and the Balance Sheet, form 16, of the business under consideration are given so as to show that all the expenses of the business are dealt with on the following forms.

When the printer has decided on the number of his departments, he can make out form 1. He must add a column for materials, to find the cost of handling, etc., and another for overhead expenses, which are explained later.

From the Trading and Profit and Loss Accounts for the last year he must find the total amount expended in:

1. Wages and National Insurance Act payments (including foremen and casual labour, but excluding office salaries, etc.).

2. Materials that can be charged to customers, eg. ink, paper, strawboards; but excluding sundry materials, such as proof paper, lye, roller composition, string, etc., which cannot be charged on each job.

3. *All* the other general expenses of the business, such as salaries, rents, taxes, insurance, interest, depreciation, etc., and sundry materials, as proof paper, roller composition, etc.

These three amounts should be placed. at the top of form 1, and the total should agree with the total expenses as shown on the Trading and Profit and Loss Accounts.

If the last year were an abnormal year for any reason, very busy or very slack, or with some unusual expenditure, take the average of the last three years as a basis for these figures. If the expenses are known to have materially altered in the year under consideration, *e.g.* rent or rates increased, or additional capital brought into the business, the figures should be adjusted to include these alterations.

This form only deals with the general expenses of the business. The expenses of the previous year or years should be entered in column 1, and divided over the different departments in the following manner:

RENT, RATES, TAXES, HEAT, LIGHT, AND WATER.- These can all be grouped together, and divided over the departments according to the square feet of floor-space in each.

Measure the floor-space in each department. The space of staircase, lifts, yards, etc., can be ignored, as they serve all departments, and their rent will be covered by the amount charged to each department. Add together the square feet in each department, and divide into the total cost in column 1. This will give the cost per foot, and the exact amount for each department can be entered in each column. The rent of office space should be placed in the overhead expenses column. In the business shown in these forms the office was in the works, and the space so small that the printer did not make a separate entry.

If you own your premises, do not forget to charge a fair rent to the business; if you do not do so, you are making a gift to your customers of the rent of the building.

If you wish to be very accurate, you can divide the cost of lighting according to the number of gas jets or electric lamps, heat according to the cubic capacity of each department, and the water according to the amount used in each department. But the square-foot principle is quite satisfactory.

POWER.- The cost of driving the machinery must be allocated to the departments in proportion to the amount used.

If you have electric motors, you can divide the horse-power of the motors in each department into the total cost. If you cannot do this, divide the annual cost over the departments by estimating the amount used in each.

DEPRECIATION.- The Committee recommends that depreciation should be written at the rate of 10 per cent. per annum off the diminishing value of the Plant. The depreciation should be allocated to the departments according to the value of plant in each.

If you have an inventory of the plant, it is easy to find the value of plant in each department. If you have no inventory, the total value of the plant as shown in last year's balance sheet must be divided amongst the departments, in accordance with the plant in each.

Remember the depreciation is not 10 per cent. per annum off the original cost of the machine, which would write a machine down to nothing in ten years; but 10 per cent. per annum off the *diminishing* value. If a machine cost £100, 10 per cent. depreciation in the first year would reduce it to £90; at the end of the second year to £81, at the end of the third year to £72 18s. 0d., and at the end of twenty years it would be reduced to about £12. On some plant, such as boilers and shafting, 10 per cent. may be rather high, but on type in constant use and fast-running machinery it is not enough. On the average 10 per cent. is correct; but if you prefer you can write a varying percentage of the value of plant in each department, according to its character, but it should not be less than 10 per cent. in the aggregate. The value of new machinery is added annually to the value of plant in each department as purchased.

FIRE INSURANCE must be allocated to each department as follows: Insurance on buildings in proportion to the floor-space, and insurance on plant and contents according to the value in the department.

It has already been explained how to obtain the floor-space and plant value, and last years stocktaking will give you the value of stock and work in hand in each department.

WORKMEN'S COMPENSATION.- The sums spent in compensation or insurance against claims should be allocated to the department on the basis of the wages paid.

REPAIRS AND RENEWALS TO PLANT AND MACHINERY.- The sums thus expended should be allocated to the departments where the expenses are incurred.

If you do not know what the expenses were in each department last year, the amounts should be estimated. In the future a record should be kept of the departments in which the expenses are incurred, so that when form 1 is made out in the following year, these details are correct. By this method the costs of repairs will be borne by the processes which necessitate these expenses.

Repairs must not be treated as an addition to the value of the plant. You are only maintaining the plant in running order, and these expenses must be treated as a working expense.

"Renewals" in this case does not mean putting in a new machine, which is an addition to plant, but renewals of parts to existing machines.

DIRECT DEPARTMENTAL SUNDRIES.- These include the various non-chargeable materials in each department, which cannot be charged to each job, such as proof paper, lye, roller composition, oil, rags, etc., etc.

If you have no records for the previous year, the amount must be estimated for each department. In future keep an analysis book of your purchases of materials, with columns for direct departmental sundries for each manufacturing department, and one or more columns for changeable materials (paper, ink, etc.).

INTEREST ON CAPITAL.- Interest at 5 per cent. per annum on the capital employed in each department should be considered as an expense of the business, and allocated to such departments.

You know the total capital in your business. If you add the value of plant to the stock and work in hand in each department, you will find the amount of capital therein, and the interest on the balance of the capital in the business should be put, in the overhead expenses column.

It is a mistake to think that interest on capital is not an expense of the business. You are in business, and are risking your capital, to obtain a nett profit. If your business increases, you will probably obtain more capital by loans from your friends or bankers, or in the form of preference shares if your business is a company ; and on this additional capital you will have to pay at least 5 per cent. interest. You will not benefit until after 5 per cent. has been earned. Besides, it is no use risking your capital in a printing business unless you can obtain more interest on your capital than if you invest in Stock Exchange securities.

Another reason for debiting interest on capital to the departments is that certain processes require more expensive plant than others. In the machine room the value of the plant when compared with the wages is usually much greater than in the warehouse; and unless each department is debited with interest on capital, you will not find the correct hourly cost for the different processes.

ODD MEN, ETC.- If you have an engineer, labourer, or a liftman, their wages should be entered on this form, and allocated to those departments where their services are rendered, and the wages of a paper warehouseman to the paper department.

As far as possible everything should be allocated to the departments, so as to reduce the overhead and increase the direct departmental expenses.

ALL OTHER GENERAL EXPENSES should be placed in the overhead expenses column, to be dealt with as a percentage on materials and departmental costs, as explained later.

Among these expenses should be a salary for yourself. When you charge a job to a customer you include the cost of the wages of your employees, the cost of materials, and a charge for the use of the machinery and buildings. Why should you not charge your customer for the services you have rendered him? Yours are probably the brightest brains in your business, and should be well paid for. You are entitled to be paid for your services just as much as your manager and foremen, in addition to obtaining a reasonable interest above 5 per cent. on the capital locked up in your business.

If in a business there is an active partner and a sleeping partner, each owning half the capital, the active partner would not be willing to draw the same interest as his sleeping partner and no salary. He would want a salary in addition to his interest, and if you are the sole proprietor, you should pay yourself a reasonable salary.

These expenses are put in the overhead expenses column, as it is impossible to say to which department they belong.

If you have customers in the town where your works are situated, and the travelling expenses are small, *e.g.* 5 per cent., and also customers in different parts of the country, with heavy travelling expenses of 15 to 20 per cent., you must deal differently with the travelling expenses. In the Overhead Expenses on form 1 do not put all your travelling expenses, but only 5 per cent. on your annual turnover to cover your minimum travelling expense. Your hourly rates will then include 5 per cent. for travelling, and your job cost sheets for customers in your town are correct; but on cost sheets for country jobs you would have to add 10 or 15 per cent., or whatever you have found to be correct, to cover the heavy expense of your country travellers, and to obtain your total cost.

If you occasionally have abnormal expenses for carriage to distant places, they should dealt with in the same way.

Each departmental column should be added up, and the totals divided by 52 will give the weekly assessment of departmental expenses in each case.

Another method is to divide by 50 to obtain the amount of the weekly assessment, and in the four holiday weeks - Christmas, Easter, Whitsuntide, and August - to charge the departments with half the usual weekly assessment.

In the business shown on these forms the weekly assessment of departmental expenses was:-

	£	s.	d.
Composing	3	2	9
Machine	2	15	6
Binding	1	0	9

These amounts appear each week on the Statement of Cost of Production (form 3), and should be maintained for twelve months, and then forms 1 and 2 should be filled up again, and the weekly assessment amended if necessary. Some printers prefer to exclude carriage from this form, and to add the cost of carriage to each job cost sheet. In this case the actual sum spent on each job should be added to the cost sheet, or a fixed sum per cwt. according to the weight of paper used on the job.

GENERAL EXPENSES ON MATERIALS.- The next point is to deal with the expenses allocated to the materials column and the overhead expenses. There are two ways of finding out the amount that should be added to materials. The first, and by far the better, method is as follows:-

The expenses of storing and handling materials are already shown in the materials column. To this column should be transferred a fair proportion (see below) of the "all other expenses" from the overhead expenses column. The increased total of the materials column will be a percentage on the total amount of the materials for the year, and this percentage should be added to the net value of all material supplied to customers to cover the costs of handling and selling.

The following is a very satisfactory method of allocating a fair proportion of overhead expenses to materials. Find out the percentage your total materials for the year bear to your total sales for the year. In this business the materials are £750, and the total sales £3,000 (*i.e.* 25 per cent. of the total sales). Transfer 25 per cent. of the "All other General Expenses" from the overhead expenses (col. 7) to materials column. 25 per cent. of £362 = £90 10s. 0d. £27 is already in the materials column, for storing, etc. The total of the materials column thus becomes £117 10s. 0d., which is 15 per cent. on the value of the material for the year. In this business, therefore, 15 per cent. would be added to the net value of material supplied to customers, to cover the cost of handling and selling.

When estimating or charging you should put down the nett cost of material, add the percentage you have found to be correct (whether 10, 15, or 20 per cent.), and you then have the total cost of material and the handling and selling expenses, and to this total you must add your nett profit.

If you supply much ink, binder's cloth, etc., it is desirable to have a separate column for these materials, as the cost of rent, rates and taxes, insurance, and carriage will probably be a smaller percentage on them than on bulky materials such as paper, of which you may have to hold a large stock.

If your customer supplies the paper to you, you must not forget to include on your estimate and job cost sheet a sum to cover the costs of handling and storing. In this case find the approximate number of reams (whether your own or your customer's) that enter your warehouse in a year, and divide this number into your annual costs of handling piper, to find a price per ream. On customer's paper you do not pay traveller's commission, and there is no interest or depreciation; but you do pay rent, etc., ware-housemen's wages, spoilage, carriage, etc. When you have found the average cost per ream for handling, you should fix that rate for medium-size reams, and more or less for larger and smaller reams.

If you decide not to adopt the above method, which is the better and more accurate way, you can fix on an arbitrary percentage to add to materials, which should be *at least* 10 per cent. to cover your costs, and should probably be more. In the business shown, 10 per cent. was adopted, which amounted to £75. £27 had already been placed in the materials column as the cost of handling and storing, and the balance of £75, viz. £48, must be deducted from the overhead expenses column.

Your handling and storing expenses on materials will very likely amount to 5 per cent. or more; your traveller's salaries, commissions, and expenses will probably be from 5 to 10 per cent.; and in addition there are other expenses on materials, such as interest, discount, carriage, bad debts, and spoilage, so that 10 per cent. will almost certainly be *less* than the amount you should add to materials.

You have now allocated all the general expenses, either to the departments, to materials, or to overhead expenses.

METHOD OF RECOVERING OVERHEAD EXPENSES (2)

This form shows how to recover the balance of overhead expenses, viz. £339, in the business under consideration. The Committee recommends this amount should be recovered by adding a percentage (which will vary with each business) to the total departmental cost, *i.e.* the wages paid plus the standing departmental expenses. Form 2 shows how this is found. To the wages paid (£1,050) add the total departmental expenses (form 1, column 6, £361), and the percentage the overhead expenses are to this total must be calculated. In this business the overhead expenses of £339 are 24 *per cent.* on the total of wages and departmental expenses. Add the overhead expenses (£339), and the total (£1,750) is the labour side of the business. Add the total value of materials and the percentage thereon, and this will show that the total expenses of the business, as shown on the top of form 1, have been recovered by these various allocations.

Forms 1 and 2 should be made out once each year at the beginning of the financial year, and the assessments varied if necessary. Your auditor can probably make out these forms without any difficulty, and he will see that all the expenses are included.

WEEKLY STATEMENT OF COST OF PRODUCTION (3)
On this form should be entered each week the wages paid in each department, and to this amount should be added the departmental expenses as shown by form 1. To the total of these two amounts add an amount representing the percentage for overhead expenses as found on form 2, and the final total is the whole cost of each department for the week. Rates for each chargeable hour (*i.e.* actual hours spent on jobs) must be ascertained, so, as to recover this amount when all jobs are charged up.

The value of production (see form 9) is the total of all the chargeable hours (*i.e.* actual hours charged on the job cost sheets) in each department, multiplied by their respective hourly rates. The difference between the total departmental cost and the value of production shows whether the department is busy or slack, and covering its costs or not, and when the results of a sufficient number of weeks are compared it shows if the hourly relates are correct or not.

The wages of the employees must be grouped in the wages book, so as to give a separate total for each department. If the overseers are paid monthly, a proportion of their salaries must be included weekly.

DEPARTMENTAL ANALYSIS (4)
This is a summary of the differences between the departmental costs and the value of production in each department, from form 3. If the hourly rates are correct, the surpluses should approximately balance the deficits over a number of busy and slack weeks.

In this business the printer had correctly ascertained his rates for machining and binding, but his composing hourly rate was too low, as there is a continued deficit.

If you find a deficit in any department, you must increase your hourly rates until the two columns approximately balance, and you then know you have found out your actual cost per hour for each process. You cannot accurately know your average hour costs for each process until you have figures for a year. Estimates should be based on average costs, not on the costs of either slack or busy seasons.

DAILY DOCKET (5)

In order to find the number of hours worked on each job in each process, a daily docket must be kept by each worker whose time is chargeable, and the docket should show the number of hours worked on each job, and its number, and the number of non-chargeable hours.

Any further information can be given on the docket, but the above information is essential.

Daily docket No. 5 is suitable for a small business, and can be used by a man who works in more than one department. If he works in more than one department, his wages must be debited to the departments according to the hours worked in each. It will probably be more convenient to have separate dockets, giving special information for each department, specimens of which are given in forms 6, 7, and 8. If a man is working on two machines, a separate docket should be made out for each. The docket for each machine includes the normal workers on the machine, *e.g.* minder and feeder, but the extra cost of a second minder helping to make ready should be added.

In some businesses a weekly docket is already in use for chargeable time. If the hours on these dockets are transferred each morning to the job cost sheets, there is no need to use daily dockets. Weekly dockets should be collected each morning, and returned to the workers as soon as the previous day's hours have been transferred. It will, however, be found much simpler and easier to have daily dockets instead of weekly dockets.

Make out dockets of any shape or kind to suit your own business, but each worker whose time is chargeable should show all the hours he has been paid for on his daily dockets. If the proprietor or his foreman occasionally works at composing or machining, he should make out a docket so that the hours he has worked a may be recorded against each job. He must also charge a corresponding proportion of his salary to the departments in which he has worked. The balance of his salary will be part of the overhead expenses.

You must not confuse "working" hours with "chargeable" hours. A man works, say, 51 hours per week, but you cannot directly charge all those hours to individual jobs. In all departments there is the time of overseers, boys, porters, and other workers, whose time cannot be charged against each job, so their time is non-chargeable.

In the composing room readers and reading boys, proof-pullers, storekeepers, and time on distribution and clearing are all classed as non-chargeable. In the machine room there is idle time - time spent in oiling-up machines, etc.; and in the binding room idle time, oiling machines, time workers waiting for work, boys who carry work, etc., must all he treated as non-chargeable. If you wash up for a special colour, the time spent should be charged to the particular job. A general wash-up at the end of the day or week is non-chargeable time.

The Committee recommends that *the compositor hours on a job should be the time spent in composing, correcting (both hose and authors'), imposing and sending to machine*, and that all other work should be considered non-chargeable. You may compose, print, deliver, and charge up a job weeks before the type is distributed, and it is impossible to charge the time spent on distribution to each particular job. Time spent on reading, proof pulling, distribution, clearing, etc., is best charged for by increasing the hourly rate of the compositor.

"Non-chargeable" does not mean the time is not charged to the customer, but that it is not charged in detail to each job. The value of non-chargeable time is included in the hourly rate for chargeable time.

If the employees are paid on piece, the daily dockets must show the wages paid on each job, and in addition the composing-room dockets must give the thousands of ens set. If the piece workers' weekly wages sheets or books are sent to the office each morning, the amount of wages earned on each job can be transferred to the job cost sheets, and a daily docket is thus obviated.

A certain amount of overtime in the printing business is due to general pressure of work, and cannot be charged to any particular customer. The cost should be covered by the hourly rates. If overtime is specially worked to meet the customer's requirements, a reasonable amount to cover the extra cost should be added to the job cost sheet.

VALUE OF PRODUCTION (9)

The daily dockets must be sent each day to the clerk who deals with them. The total chargeable hours in each department must be entered on the Value of Production Form (9). At the end of the week the chargeable hours or piece units of each process must be totalled and multiplied by the determined rates, or the piece wages totalled and a percentage added. The methods for finding these rates are fully explained under Hourly Rates. The results give the value of production in each department, and must be entered on the Statement of Cost of Production (3).

In a small business like this example all the departments can be kept on a single sheet; but in a larger business it will be more convenient to have separate sheets for each department.

In this business the compositor hour was reckoned at 1s. 6d. (Form 4 shows this rate was incorrect.) In the machine room each machine should be numbered, and the chargeable hours entered in its own column. The hours of each machine should be multiplied by its own hourly rate. In the binding room the time workers were reckoned as follows: women on hand work at 6d. per hour and on a wiring machine at 9d. per hour; men on hand work at 1s. per hour; men on a cutting machine at 1s. 6d. per hour; and 100 per cent. was added to piece-work wages of men an women. These rates are not given as standard rates, but as an example in this business. Your rates may be higher or lower according to the circumstances of your business.

In a larger business there would probably be more hourly rates, as men at 40s. per week must be charged at higher rates than those at 30s. per week, and there would be different rates for machines of varying values.

The machine hourly rates exclude the value of ink, which is considered as chargeable material. A machine may be running on common work with ink at 9d. per lb., or on a high-class three-colour half-tone job with ink at 5s. per lb. The overseer can put the value of the ink used on the daily dockets or on the work ticket of the job, and this amount must be added to the Job Cost Sheet (10).

If the work is done on piece, the amounts paid must be entered instead of the number of hours, and at the end of the week a percentage (found by experience to be correct) should be added to cover general expenses.

JOB COST SHEET (10)

As soon as an order is received, this sheet should be made out and kept in a loose leaf binder in the office. Each day the number of chargeable hours shown on the daily dockets should be transferred to these sheets. Chargeable materials should be entered on the sheet when issued. When the job is finished, the chargeable hours and piece units must be multiplied by the correct rates, and the percentages added to the cost of chargeable materials. The totals from all departments give the cost of the job, including a proper proportion of all general expenses of every kind. To this amount the printer should add a reasonable amount for *nett* profit, or if he has given a price for the job the difference between the cost and the price given is the nett profit. It should be remembered that if the nett profit of a job is 10 per cent. and 2 1/2 per cent. discount is allowed, you are giving away 25 per cent. of your nett profit.

The job cost sheets are much more convenient when kept in a loose leaf binder. When a job is finished, its cost sheet can be detached an (filed away with its work ticket (12), and a file copy of the job, and all the details are readily available for reference. The cost clerk only has to handle "live" sheets, when posting from the daily dockets.

If there are many departments, or the job takes a long time, the departments can also be printed on the back of the sheet, or separate sheets for each department can be inserted behind the original sheet.

Further details can be added to the cost sheet, such as employee's number, overtime hours, etc.; but these details must be settled by each printer.

In this business the binding-room hourly rates were women on time at 6d., women on wiring machines at 9d.; men on time at 1s., and men on cutting machines at 1s. 6d. Columns must be added at varying rates according to the different machine and time-workers' rates. If the work is done on piece, the amount of piece-work wages must be entered in a special column. The total paid, plus a percentage (see Binding-room Hourly Rates), will give the total cost.

Outwork includes manufactured goods bought in for the job. The percentage to add for handling and selling expenses will be less than on paper, as there is no interest or depreciation on stock, etc.

ORDER BOOK (11)

As soon as an order is received, it should be entered in this book, and given a number which should be repeated on all work tickets, cost sheets, and daily dockets. When the job is charged up, the folio of the day book should be entered in the outer column. This will prevent any work being delivered and not charged.

WORK TICKET (12)

A work ticket in the form of an envelope is convenient for holding copy, patterns, etc., and should accompany the job through the Works; and the job cost sheet should be inserted when the job is finished. This ticket will contain the whole history of the job, list of revises, deliveries, etc., the cost and the nett profit. When the order is repeated, the former work ticket should be referred to, to ascertain if the job paid, and what extras can be charged for.

In a very small business where everything is under the proprietor's eye, it is still desirable to have a work ticket, although it may not leave his desk in the Works. All the instructions are recorded on it, and extras can more easily be traced, and if desirable the Job Cost Sheet (10) can be printed on the back.

This work ticket is only a suggestion of a simple form. Each printer must draw up a work ticket to suit his own requirements.

The time taken by a clerk to transfer the hours from the daily dockets to the Value of Production and Job Cost Sheets is not very long. Printers who have installed the system find it takes from three to five minutes daily for each docket. The time depends upon the number of entries on each docket. Probably only two-thirds of your employees will write dockets, the remaining third will be non-chargeable, *e.g.* overseers, readers, porters, etc., or their time will be covered by a machine docket, *e.g.* layers-on on printing machines. Probably every printer has some form of daily or weekly docket in use, which can be adapted to the system so that the cost of the extra work of transferring the number of hours can easily be ascertained.

HOURLY RATES AND MACHINE UNITS

When a cost-keeping system is first installed in a printing office, it is impossible to begin with hourly rates that are certain to be correct. The actual hourly rates can only be found by experience.

Probably every printer has some hourly rates in use for each process in his Works, and these can be continued until the Departmental Analysis (4) proves whether these rates are correct or not.

It cannot be too strongly impressed on the printer that he should estimate and put on his job cost sheet the hourly rates found to be correct by the departmental analysis. If he puts lower rates, he is deceiving himself, and may be making losses on jobs which appear to show a profit.

The rates for each department can, however, be fixed in the following manner.

COMPOSING

If the department works on time, the total chargeable hours shown on the Value of Production (9) should be divided into the total departmental cost shown on 3. If the department is doing piece work only, the thousands of ens set should be divided into the total departmental cost. The result is the cost per hour on time or per thousand ens on piece.

As an example, let us suppose that at the end of three months you wish to check the hourly rates you are using. You must total the chargeable hours on the Value of Production (9),-and the total departmental cost on the Statement of Cost of Production (3) for the three months, and divide the former into the latter, to find the correct hourly cost. If the total chargeable hours were 2,500, and the total departmental costs £218 15s., you would find your compositor hour was costing you 1s. 9d.

The hourly rates cannot be accurately ascertained in less than twelve months, so as to cover slack and busy seasons, and the cost will vary considerably at different times in the year. Your rates for estimating and charging should not vary. The Departmental Analysis (4), when totalled for a long period, will show if your rates are correct.

Do not be alarmed if your compositor hour on time comes out at more than you expected. A Chartered Accountant in 1904 examined the figures of a large number of printing offices in London. He found that the compositor hour, including distribution, cost about 1s. 11d. This was when wages were 39s. per week of 52 1/2 hours. Your cost per hour will probably be about the same, subject to adjustments for different rates of wages and the peculiarities of your own business.

If the forme for a job is standing, this should be noted on the daily docket and on the job cost sheet. If possible, the job should be recharged to your customer as though reset. Your type and material have been locked up, and you have run the risk of the job not being repeated. Any advantage should belong to you and not to the customer.

Apprentices in the composing room can be charged at a rate according to their year of service - *e.g.* one-third journeymen's rate for the first three years, two-thirds for the next two years, and full rates for the last years of their apprenticeship. It would be well to revise these rates each year, as apprentices vary in efficiency. Another method is for the overseer to mark on the apprentice's daily docket the hours a journeyman would have taken, and to enter these hours on the job cost sheet and value of production form.

If there are monotype or linotype machines, it is necessary to divide the composing room into two or more departments. The Statement of Expenses (1) will give the weekly standing departmental expenses for the machine- and hand-composition departments. On the Statement of Cost of Production (3) the wages paid to employees in the machine department and to compositors on time will be shown separately. The overseer, readers, proof pullers, and storekeeper are all working both for the hand and machine departments, and their wages must be divided proportionately between the two departments.

A satisfactory method is to reckon that one thousand ens, set and corrected on the galley, costs the same amount in oversight, reading, etc., as an hour's composition on time. The wages of the overseer, readers, etc., must then be entered on the Statement of Cost of Production (3) in proportion to the chargeable hours of time composition, and the thousands of ens. produced on machine in the week.

The total of the wages paid, and the share of overseer and readers' wages, and the departmental expenses, gives the direct departmental cost in the hand- and machine-composing departments, to which should be added the percentage for overhead expenses, and the result is the total departmental cost.

If you prefer, you can calculate the production of the machine composition by the number of hours worked on the keyboards, and reckon that each keyboard hour is equal to five to eight compositors on time according to your machine output.

If you compose and correct matter on galleys in the machine-composing department and make up in the hand department, your costs will appear on your job cost sheet as a certain number of thousands at the rate you have found for machine composition, and a certain number of hours' hand work at a higher rate.

If the production of your machine and hand departments is uniform, you can divide the costs of overseer, readers, etc., on a fixed basis which experience shows is correct; but it is better to divide these wages weekly, as the output may vary considerably, and then your reading expenses would not be fairly allocated.

FOUNDRY

The output should be reckoned in square inches at different, rates for stereos and electros. The value of the inches produced for each job should be entered on the job cost sheets, and the value of the total inches for the week should be shown on the Value of Production (9).

These rates do not include the value of metal in the plates. If the stereos or electros are sold to your customer the value of the metal must be added to the cost of making.

If you have no rates, you can take the rates charged by the trade electrotypers in your town, less, the value of metal in the plates, and use those rates for your own costing, until you find your actual cost.

PRINTING MACHINE UNITS (FORM No. 13)

If you do not wish to work out the cost of each machine by the unit system, you can take the hourly rates for machines as shown in "Profit for Printers" (published by the Federation of Master Printers), and the weekly statement of cost of production will soon show if these rates are correct or not for your business.

If you wish to find out your own rates, the following method should be adopted.

A unit representing machine hour cost must be found for each machine. This unit will cover all the expenses of the department-with the exception of the wages of the minder and the feeder on each machine-and to the unit hour cost of the machine a sum must be added to cover the wages of minder and feeder. The ink used should be excluded from the hourly rate of the machine, and the value of ink used on each job plus a percentage for handling and selling expenses should be entered on the job cost sheet.

Each printer should fill up a form similar to No. 13 for his own business. (This form is filled with details from a different business to that shown on the previous forms.) The first column should contain all the machines, and column 2 the original cost of the machine.

The Committee consider that the original cost of the machine is a fair basis for allocating all the expenses of the department, as depreciation, insurance, and interest are according to the cost of the machine; and floor-space, power, and overhead expenses, such as traveller's commission, carriage, etc., approximately follow the cost of the machine.

It is better to take the original cost, and not the depreciated value, of the machines. If you work out your hourly costs on the depreciated value of your machines, and if all the machines were depreciated equally, the result would be the same, as you would have fewer units to divide into your costs, and the unit value would, therefore, be higher; but as your machines are probably of varying ages, the depreciated value of similar machines would vary, and consequently you would have varying hourly rates for similar machines. If you have some very old and obsolete machines, running very few hours in the week, it might be better to fix the number of units according to the output of the machine instead of on the £50 basis.

Each machine must be reckoned as representing a certain number of units. It is suggested that £50 is a convenient unit. Column 3 shows the number of units for each machine, assuming one unit for every £50, or part of £50, of the original cost of the machine. Column 4 shows the ascertained average number of chargeable hours of each machine in a week.

You are paying rent, rates and taxes, interest on capital, and an overseer's salary for 51 hours or more per week, but your machines will probably only run an average of 35 to 45 hours per week all the year, so it is useless to divide your weekly costs by 51 hours, if that be your working week. You must divide your costs by the hours the machine is running, which is the time you can charge your customer. The average number of chargeable hours is shown in column 4. The number of hours should be ascertained over as long a period as possible.

Having ascertained the number of units the machine is assessed at, and the average number of chargeable hours for such machine, multiply them together, and you will obtain the total hour units for each machine per week. These hour units are placed in column 5, and when the hour units for all machines are added, you have the total for the whole machine room. In this business there are 1,318 hour units per week.

To find the value of the hourly unit you must take the total departmental cost of the machine room for as many weeks as possible, and find the average weekly cost. In the business shown, the average weekly cost for the department was £22 15s. 6d. From this sum deduct the average weekly chargeable wages (*i.e.* the wages paid to minders and feeders), and the balance is that part of the cost of the machine room which must be covered by the machine unit hour cost. In the business shown this machine cost was £14, which divided by the total hour units (1,318) gave an hourly unit value of 2.55d. In column 6 the units of each machine (column 3) multiplied by the hourly unit value (2.55d.) give a machine unit hour cost for each type of machine.

It is not necessary to take £50 as the unit of value of machine. You can fix this at any amount you choose. The hourly unit value will not be the same in your business as in the example shows. It will be more or less according to the cost of your machines, and whether you adopt a unit of more or less than £50. It will also depend on the costs of your department. You must find out the units of your machines and your own hourly unit value before you can fill up column 6. Column 6 is the result of all the previous columns.

The wages paid to feeders on each machine should be entered in column 7, and to minders and apprentices in column 8.

The wages and particulars on this form are taken from a provincial town. One minder at 35s. 6d. with two feeders is working on a royal and a demy wharfedale. The total wages paid to feeders are £2 7s. 6d. per week; minders and apprentices £6 8s.; total chargeable wages £18 15s, 6d.

An apprentice earning 10s. per week was working one quad-crown wharfedale, and if the rate for that machine were worked out on these wages, it would be less than similar machines worked by a journeyman. To equalise the lower cost of machines with apprentices, it is necessary to proceed as follows. In column 9 place the wages that would be paid if there were a journeyman on each machine.

In this town 35s. 6d. was the rate for a journeyman on a quad-crown wharfedale. On one of the machines 37s. 6d. was paid for a minder doing better-class work.

The total of column 9 must be reduced by one-sixth to bring it to the total of wages actually paid. Column 10 shows the journeyman's rate of wages reduced by one-sixth in each case, so that the total is the same as the wages actually paid for journeymen and apprentices.

By this adjustment the advantage of the lower wages paid to the apprentices is not taken by any particular machine or job, but is spread over the whole department. Many printers will prefer to work out machine rates as if a minder were on each machine, as the apprentice is more likely to spoil work, and time and attention is taken in instructing him.

The total chargeable wages paid on each machine (after the adjustment to include the apprentices at lower wages) in column 11 is found by adding column 7 to column 10. The wages (column 11) must be divided by the average chargeable hours of the machines (column 4) in order to find the wages per chargeable hour (column 12). The chargeable wages per hour (column 12) should be added to the machine unit hour cost (column 6) to obtain the machine hourly rate (column 13).

If your work is seasonal, and your staff increases and decreases accordingly, the average chargeable hours per week of the minders and feeders may be higher than the hours of the machines. Thus the machines may only average 40 hours per week throughout the year, but the minders and feeder may average 42 or more during the time they are engaged.

You will notice that the wages paid to minders on the two quad-crown machines are different. It is desirable to have a uniform rate for the same type of machines. Therefore, the total wages paid on both the quad-crown machines are added together and divided by two to find the average wages paid on the quad-crown machines. This gives a uniform hourly rate for the wages.

As there are one minder and two feeders on the royal and demy wharfedales, the total wages paid must he added together and divided by two to find the rate for each machine.

Your minders and feeders are paid for a full week, but you cannot charge the whole of their wages to different jobs, as time is lost by idle time, oiling up on machine, waiting for work, breakdowns, etc., etc., so the wages paid per week must be divided by the average number of hours per week that you can charge their time.

If an extra minder assists in making ready on a machine, his daily docket should show the hours and the job on which he was helping. These hours should be added to the job cost sheet, at a rate to cover the wages paid, plus a percentage for supervision, etc.

If a minder is working on two machines, he must write a daily docket for each machine.

The hours spent on each job will be found from the daily dockets, and these hours should be multiplied by the correct machine hourly rate (column 13) to find the cost. To the machine cost must be added the cost of ink used, plus the proper percentage for handling and selling materials.

If you only do ordinary jobbing work, with ink at a uniform price, you can work out the average cost per hour of ink for ordinary jobs. This hourly cost of ink should be added to the machine hourly rate. Whenever a job is printed which uses more expensive ink, or an abnormal quantity per hour, the value of ink used should be found and the extra cost above the few pence per hour, which is already included in the machine hourly rate, should be added to the job cost sheet.

In estimating and charging for machine work, the time of making ready and running on should be reckoned at the machine hourly rate. You only save a very small portion of the power when the machine is making ready. It is not worth while having separate rates for making ready and running on.

In estimating for work it is unwise always to reckon the hourly rate of a machine of the exact size of the sheet. If all double-crown jobs are estimated at double-crown machine rates, you will probably find you many of them on double-demy or larger machines. It is better to estimate for a machine slightly larger than the job, so as to be certain to cover your costs. Your customer will not usually wait until you have the correct machine at liberty. If you are very slack, and put a double-crown job on a quad-demy machine, you cannot expect your customer to pay for this machine. The machine number on the job cost sheet will show that the cost was abnormal.

When you have ascertained the hourly rates by this form, they should be continued for twelve months, or until the Departmental Analysis (4) has shown them to be incorrect after many weeks, and the form can then be made out again to ascertain what alteration is necessary.

WAREHOUSE OR BINDING ROOM

In the machine room all the costs are recovered by charges for machine work, but in the binding room a large portion of the costs are recovered by charges for hand labour, and therefore a different system must be adopted.

Machines which cost less than £10 should be treated as tools, and their cost be recovered by the hourly rates for hand work. Machines of greater value should be entered on a form similar to No. 14.

Put each class of machine in a separate column, showing the original cost of the machines, and the floor space they occupy. In each column put the cost of depreciation, rent, power, etc., and the amount spent in repairs, knife grinding, etc. The total of each column, divided by the number of machines and again by 52, gives the weekly cost of each machine; and if this is divided by the average chargeable hours per week you obtain the net hour cost of each machine.

To find the total cost per hour, add the hour cost of the machine to the wages paid per hour to the workman. To this total add the ascertained percentage to cover the balance of the costs of the department, and the result is the total cost per hour for each machine.

As an example, let us suppose that your cutting machine cost is 4d. per chargeable hour. If your cutters are paid 8d. per hour, your nett hour cost is 1s. To this you must add the ascertained percentage to cover all the other expenses of the department, say 100 per. cent., or whatever percentage experience proves to be correct, and this would make your rate for a cutting machine 2s. per hour.

For hand work, if the man works on time at 8d. per hour, you must add say 100 per cent. to his time wages, and his rate would be 1s. 4d. per hour.

In the same way, a girl whose time money is 3 1/2d. per hour-if working a sewing machine that you knew by form 14 cost 4 1/2d. per hour = 8d.- would be rated at 1s. 4d. per hour.

If the employee is paid piece work on the machine, you must find the average amount earned per hour. Suppose the piece worker earns 9d. per hour on a machine, and the machine rate is 4d. Then for every 9d. paid in wages on the job you must add 4d. to cover the cost of the machine, and add the usual percentage for expenses. The percentage you add to wages will cover the cost of idle time, nonchargeable wages, rent, insurance, and all other expenses. As far as possible every worker should make out a daily docket, so as to charge their time to actual jobs.

If you work on piece, the job cost sheet must have columns for piece-work wages paid, as well as for time at varying hourly rates. The amount of piece wages paid will be transferred each day either from the daily dockets, or from the weekly wages sheet of the piece worker, to each job cost sheet.

In a few weeks you will find out from the Statement of Cost of Production (3) if your percentages are correct, and if incorrect they must be modified accordingly.

If a special machine, such as a book-sewing machine, is installed, and at first is only partly employed, say one day a week, you cannot charge the full cost of the machine as shown by form 14. If you did, your charge per hour would be so high that you would not obtain extra work for the machine. In this case you must find the hourly cost on the assumption that the machine is running a normal number of hours, say 30 or 40 hours per week. You will then find a reasonable hourly rate for the machine, and you will hope very shortly to obtain work to keep the machine running a normal number of hours.

(*The Printers' Cost-Finding System*, 2nd ed., July 1913, pp. 2-22).

*

SPECIMEN FORMS

STATEMENT OF EXPENSES

Shewing Standing Departmental Expenses per Annum and per Week.

The following figures are hypothetical but approximately correct for the class of business shewn.

The business dealt with in this example spends **£1,050 per annum in Wages.**

750 „ „ **in Materials.**
775 „ „ **in Expenses.**

Capital Invested, £2,000.
Total sales - - £3,000.

£2,575

ASSESSMENT OF TOTAL EXPENSES BASED ON PREVIOUS YEAR OR YEARS.		DIRECT DEPARTMENTAL EXPENSES.				TOTAL DIRECT DE-PARTMENTAL EXPENSES.	OVERHEAD EXPENSES NOT DIRECTLY CHARGEABLE TO DEPTS.
		COMPOSING ROOM.	MACHINE ROOM.	BINDERS.	PAPER OR MATERIALS STOCK ROOM.		
£		£	£	£	£	£	£
130	Rent, Rates and Taxes } Light, Heat and Water } .. .. *(divided according to floor space)*	60	40	20	10	130	—
20	Power .. .. .. *(divided according to amount used)*	—	16	4	—	20	—
120	Depreciation at 10 % .. .. *(on value of plant in each department)*	60	45	15	—	120	—
10	Fire Insurance .. .. .. *(according to floor space and contents)*	4	3	1	2	10	—
2·10	Workmen's Compensation ..	1	1	·10	—	2·10	—
20	Repairs and Renewals to Plant and Machinery	6	12	2	—	20	—
10·10	Direct Departmental Sundries	2	4·10	4	—	10·10	—
100	Interest at 5 % on £2,000	30	22·10	7·10	15	75	25
362	All other General Expenses .. *(e.g., salary for Proprietor, travellers' commissions, office expenses, stationery, postages, discounts, bad debts, spoilage and general trade expenses, etc., etc.)*	—	—	—	—	—	362
775	Yearly Assessment of Standing Departmental Expenses	163	144	54	27	388	387
	Weekly Assessment of Standing Departmental Expenses, which appears in Weekly Statement of Cost of Production	£3 : 2 : 9	£2 : 15 : 6	£1 : 0 : 9	—	775	
	Deduct £75 chargeable to Materials (see note below) ..					27	48
						£361	£339

EXPENSES CHARGEABLE TO MATERIALS.

The suggestion of the Cost Committee is that each Printer should ascertain for himself the amount chargeable to Materials for Storing, Handling, Commission, Carriage, Bad Debts, &c., but that a minimum of not less than 10 % on the value of Materials used should be charged.

In this business the printer did not allocate overhead expenses to materials, but considered 10% was sufficient to add to materials. 10 per cent. on £750 Materials = £75, of which £27 is shewn to be the actual handling expense (see Col. 5) and deducted from Direct Departmental Expenses (Col. 6), the balance required of £48 is deducted from Overhead Expenses (Col. 7). (In a larger business the paper warehouseman's or a share of porter's wages would be placed in the materials column as a separate amount.)

For method of dealing with the above figures see Form 2 and 3.

This form should be made out once each year.

METHOD OF RECOVERING OVERHEAD EXPENSES.

Method of ascertaining Percentage on Direct Departmental Expenses plus Wages to recover Overhead Expenses.

The Printer having *ascertained* his Direct Departmental Expenses as explained by Form 1, should then *recover* his Total Overhead Expenses (*i.e.*, £387), (*a*) by a percentage on Materials, (*b*) and by a percentage on Direct Departmental Expenses plus Wages actually paid, as follows :—

Total Wages estimated on last year (or by taking the average of the last three years)		£1,050
Add Direct Departmental Expenses (Form 1, Col. 6)		361
Direct Departmental Cost		1,411
Add Overhead Expenses recovered on Departmental Expenses plus Wages (Form 1, Col. 7)		339
amounting to £339 = <u>24%</u> on £1,411	(*)	1,750
Materials	...£750	
Add 10% thereon	... 75	
		825
Total Estimated Expenditure for the Year		£2,575

(*) Rates for each chargeable hour (i.e., actual hours spent on jobs) must be ascertained to recover this amount. Some hourly rates are in use in every office. These should be continued (or new rates fixed as explained in this pamphlet) until they are shown to be incorrect by Statement of Cost of Production, Form 3, and Departmental Analysis, 4.

This form should be made out once each year.

STATEMENT OF COST OF PRODUCTION

For week ending November 22nd, 1912.

Chargeable Materials (*I.e.*, Ink, Paper, &c.) are omitted.

	COMPOSING.			MACHINE.			BINDERS.		
	£	*s.*	*d.*	£	*s.*	*d.*	£	*s.*	*d.*
Wages actually paid (and Insurance Act—Employers Contributions)	10	5	0	7	6	0	2	13	0
Add Departmental Expenses ... ... (as assessed, see Form 1)	3	2	9	2	15	6	1	0	9
Direct Departmental Cost ... ... (without Overhead Expenses)	13	7	9	10	1	6	3	13	9
Add Proportion of Overhead Expenses ... ... *i.e., 36% on Direct Departmental Cost. (See Form 2)*	3	4	3	2	8	4	0	17	8
Total Departmental Cost for the week	16	12	0	12	9	10	4	11	5
Value of Production in each Department ... . (as hourly or piece unit rates or at a percentage on wages obtained from Value of Production, form 9.)	13	13	0	12	1	6	4	18	6

This form should be bound as a book, and filled up weekly.

DEPARTMENTAL ANALYSIS.

WEEK ENDING.	COMPOSING.		MACHINE.		BINDERS.	
	SURPLUS.	DEFICIT.	SURPLUS.	DEFICIT.	SURPLUS.	DEFICIT.
	£ s. d.	£ s. d.	£ s. d.	£ s. d.	£ s. d.	£ s. d.
October 4th		0 17 6	0 9 6	—	0 8 0	—
„ 11th		1 4 0	—	1 12 0	0 12 0	—
„ 18th		1 9 0	2 0 0	—	—	0 10 3
„ 25th		1 15 0	0 10 0	—	0 14 0	—
November 1st		0 15 0	1 10 0	—	—	0 15 0
„ 8th		2 0 0	0 12 0	—	0 10 0	—
„ 15th		1 10 0	—	4 0 0	—	0 12 6
„ 22nd		2 19 0	—	0 8 4	0 7 1	—
„ 29th						
December 6th						
„ 13th						
„ 20th						
„ 27th						

This form should be kept in a book and filled up weekly. It shows the difference between the Total Departmental Cost and the Value of Production on Form 3.

This form shows that in this particular business the hourly rates in the machine and binding departments were correct, but the composing rate was too low, as there was a continuous deficit.

Form No. 5.

DAILY DOCKET.

Employee's Name ..Davidson.. No. ..1.......... Date ..21|11|.......... 19 1 2

Job No	Customer's Name	Description of Job	Machine Number	No. of Process	Hours Composing Comp.	Author's Corr.	Hours on Machine	Value of Ink used	Hours on Binding	Piece Wages on Binding
										s. d.
233	Michael	Circ		1	1					
237	Staines + Son	4 pp list		4		1				
238	Thomas	Poster	2	5			½			
				6						
234	Colston	Ledger	2	5			1	1/3		
				6			1			
340	Bath + Jones	Circ		1	1½		1	3ᵈ		
				2+3	½					
230	Hulcott	Price List		4		2				
	TOTAL CHARGEABLE				3	3	3½			
Non-Chargeable:				22			½			
				16	1					
	TOTAL NON-CHARGEABLE				1		½			

CHARGEABLE TIME.

COMPOSING.
1. Composing.
2. House Corrections.
3. Imposing.
4. Author's Corrections.

(If forme is standing, write "Standing" opposite Job No.)

MACHINE.
5. Making Ready.
6. Running On.
7. Washing up for Colour.
8. Waiting for Customer to pass proof.

BINDING.
9. Folding.
10. Gather and Collate.
11. Sewing.
12. Wire Stitching.
13. Cutting.
14. Binding.
15. Packing.

NON-CHARGEABLE TIME AND CAUSE OF DELAYS.

16. Distribution and Clearing.
17. Reading.
18. Proof Pulling.
19. Picking for sorts.
20. Waiting Instructions.

21. General Wash-up.
22. Oiling Machine.
23. Waiting for forme.
24.   „   „   „ paper.
25.   „   „   „ instructions.
26.   „   „   „ faulty composition or plates.
27.   „   „   „ work.

28. Waiting for work.
29. Oiling Machines.

Total hours shown must agree with the total hours of time paid for.

This form must be filled up daily by each worker whose time is chargeable, and the number of hours worked on each Job must be entered daily on the Job Cost Sheets, Form 10; and the total number of chargeable hours on the Value of Production, Form 9.

This form is a combination docket suitable for small businesses and can be used by a worker in any of the three departments, and also if he works in more than one.

This form should be adapted in size, shape and details, to suit each business. Most Printers will find it is more convenient to have a separate docket with varying details for each department.

Form No. 6.

COMPOSITOR'S DAILY DOCKET.

Compositor's Name. _Kennick_ .... No. _2_ .... Date _Nov. 21_ ...... _1912._

Job No.	Customer's Name.	Description of Job.	No. of Process.	Piece Work. No. of thousands of ens.	Ordinary Time. Comp.	Ordinary Time. Author's Corr.	Overtime. Comp.	Overtime. Author's Corr.
232	Barley	Leaflet	1		2			
235	Brett & Co	Card	4			½		
239	Parsons	8 pp. List standing	4					
243	Jones Bros	Sale List	1		2		1½	
		TOTAL CHARGEABLE			4	1½	1½	

NON-CHARGEABLE.

			5		1			
			6		2			
			7		1			
Examined by _DC_		TOTAL NON-CHARGEABLE			4			

Chargeable Time
and
No. of Processes.

1. Composing.
2. First Proof.
3. Imposing.
4. Author's Corrections.
   (If forme is standing, write "Standing"
   opposite Job No.)

Non-Chargeable Time
and
Cause of Delays.

5. Distribution and Clearing.
6. Reading.
7. Proof Pulling.
8. Picking for Sorts.
9. Waiting Instructions.

Total hours shown must agree with the total hours of time paid for.

This form must be filled up daily by each worker whose time is chargeable, and the number of hours worked on each Job must be entered daily on the Job Cost Sheets, Form 10; and the total number of chargeable hours on the Value of Production, Form 9.

Form No. 7.

LETTERPRESS MACHINES—DAILY DOCKET.

Minder's Name.... *Paterson* .... No. .....6..... Date.... *nov 21* ......19 1 2 .

Job No.	Customer's Name.	Description of Job.	No. of Runs.	No. of Formes.	Machine No.	No. of Process.	Hours. Ord.	Hours. O.T.
244	Sykes	Leaflets	2000	1	1	10	1	
						11	2½	
247	Barton	Poster	500	1	1	12	1	
						10	1	
						11	1	
220	Baker	Price List 8 pp		1	1	10	1½	1
					TOTAL CHARGEABLE		8	1

NON-CHARGEABLE.			
	14	¼	
	17	¼	
	20	½	
Examined by *B. J.*	TOTAL NON-CHARGEABLE	1	

Chargeable Time and No. of Processes.	Non-Chargeable Time and Cause of Delays.
10. Making Ready.	14. General Wash-up.
11. Running on.	15. Oiling Machine.
12. Washing up for Colour.	16. Waiting for forme.
13. Waiting for Customers to pass proof.	17. „ „ paper.
	18. „ „ instructions.
	19. „ „ faulty composition or plates.
	20. „ „ work

Total hours shown must agree with the total hours of time paid for.

This form must be filled up daily by each worker whose time is chargeable, and the number of hours worked on each Job must be entered daily on the Job Cost Sheets, Form 10; and the total number of chargeable hours on the Value of Production, Form 9. This form should be adapted in size, shape and details, to suit each business.

Form No. 3.

WAREHOUSE OR BINDING DAILY DOCKET.

Name *Bertha Martin*   No. 16   Date *Nov. 21*   19 1 2.

Job No.	Customer's Name.	Description of Job.	Quantity.	No. of Process.	Time. Hrs.	Time. O.T.	Piece Work Rate per 1,000 s.	Piece Work Rate per 1,000 d.	Piece Wages Earned. s.	Piece Wages Earned. d.
219	Barclay	8 pp list	1000	21				7½		7½
216	Smith + Son	16 pp. list	500	23	2					
215	Sutton	note books	750	22	1½					
227	Piper + Co.	Ledger	1	21	½					
214	Stevenson	Bazaar Programmes	4250	24	3	2				
			TOTAL CHARGEABLE		7	2				7½

NON-CHARGEABLE.

Examined by................................   TOTAL NON-CHARGEABLE

	Chargeable Time and No. of Processes.		Non-Chargeable Time and Cause of Delays.
21. Folding.	25. Cutting.	28. Waiting for Work.	
22. Gather and Collate.	26. Binding.	29. Oiling Machines.	
23. Sewing.	27. Finishing.		
24. Wire Stitching.			

Total hours shown must agree with the total hours of time paid for.

This form must be filled up daily by each worker whose time is chargeable, and the number of hours worked or piecework wages earned on each job must be entered daily on the Job Cost Sheets, Form 10; and the total number of chargeable hours, and amount of piecework wages earned on the Value of Production, Form 9. This form should be adapted in size, shape, and details, to suit each business.

Form No. 9.

VALUE OF PRODUCTION
FOR ALL DEPARTMENTS for week ending November 22 - 1912.

	COMPOSING.														MACHINE.							BINDERS.											Piece Wages Paid.		
	Number of Chargeable Hours.						Distribution and other Non-chargeable Hours.								Chargeable Hours on Machines Numbered							Chargeable Hours of Time Workers at Hourly Rates												
	Sat.	Mon.	Tu.	Wed.	Th.	Fri.	Total	Sat.	Mon.	Tu.	Wed.	Th.	Fri.	Total		1	2	3	4				Sat.	Mon.	Tu.	Wed.	Th.	Fri.	Total	£	s	d		£	s	d
1	4	7	6	7	6	8	38	1	3	3	2	4	2	15	Sat.	5	5	5	4	16	6	-	5	9	-	4	18	9	7	15	1	1	0		
2	5	7	6	8	7	7	40		3	4	2	4	2	15	Mon.	9	7	8	7		9	5	4	-	-	5	10	24	18	0	100%	1	1	0		
3	4	6	7	7	8	8	40	1	4	3	4	2	3	17	Tue.	11	8	10	10				5	8	6	5	6	9	39	1	19	0		1	2	0
4	3	7	6	6	7	8	37	2		3	4	2	2	13	Wed.	13	10	9						3	1	2	1	7	10	6				
5	5	8	7	7			27		3	3	2			8	Thurs	8	7	9										3	16	6			
							182							68	Fri.	7	7	10½										1	2	0			
															53	44	41	28½										4	18	6			

Total Hours
for week 182
@ 1/6 = £13·13·0

53 @ 2/- £5·6·0
44 - 1/6- 3·6·0
41 - 1/- 2·1·0
28½ - 1/- 1·8·6
Total for week £12·1·6

Total for week

This form should be kept in a book and filled up from the Daily Dockets. The total for the week of each department should be transferred to the Statement of Cost of Production Form. 9. If additional columns are needed, the form can be printed oblong or separate forms kept for each department.

In this business time workers were reckoned as follows :—Women on hand work at 8d. per hour, on wiring machines at 9d. per hour ; men on hand work at 1/- per hour, on cutting machines at 1/6 per hour ; and 100% was added to piece work wages.

Form No. 10.

JOB COST SHEET.

Customer's Name   Green & Co   Job No.   415

Description   10,000 Balance Sheets

COMPOSING COSTS.						BINDING COSTS.								
Date	Comp.	Author's Corr.	£	s.	d.	Date	@ 6d.	@ 9d.	@ 1/-	@ 1/6d.	Date	£	s.	d.
10|11|12	10					16|11|12	3	1	½		16|11|12	2	0	
11 "		2				17 "	3	½	½		17 "	2	6	
13 "		1					1					2	0	
14 "		1					7	1½	1			6	6	
	10	4								100%	6	6	
	4									7 @ 6d.	3	6	
	14		1	1	0					1½ @ 1/-	1	6	
										1 @ 1/6	1	6	
(Hours reckoned at 1/6											19	6	

(Hours reckoned at 1/6
This rate is shown to be
incorrect by form No. 4)

MACHINE-ROOM COSTS.						PAPER OR MATERIALS.					
Date	Mach. No.	Hours	£	s.	d.	Date	Particulars	@	£	s.	d.
15|11|12	3	1				15|11|12	10 Rms 12 2s				
16 "	"	12					C R L Post	4/	2	2	6
		13	1	6	0		15%			6	4
(Machine No. 1									2	8	10
is rated at 2/- hr.)											
Ink			1	4							
		15%		2							
		1	7	6							

OUTDOOR COSTS.							SUMMARY.		£	s.	d.
Date	Order No.	GOODS.	@	£	s.	d.	Composing		1	1	0
14|11|12	2593	10,000					Machine		1	7	6
		Envelopes	6/-	3	.	.	Binding			19	6
			10%		6	.	Paper or Materials		2	8	10
				3	6	.	Outdoor Costs		3	6	0
							Total Cost ...		9	2	10
							Add for net profit % ...		1	18	2
							(An estimate) Total Charge ...		11	1	0

In this particular business, the hourly rates in the binding room were :—Women, 6d. ; men, 1/-, and men on cutting machines, 1/6 per hour. The number of columns and rates must be adjusted to each business.
This form should be kept in the office in some form of loose leaf binder, and the costs entered on it daily from the Daily Dockets. When the job is finished, this sheet should be attached to and filed away with the Work Ticket 12. For very large jobs extra sheets suitably printed for each department can be added behind this form, on which the costs are collected.

Form No. 11.

ORDER BOOK.

Date.	Work Ticket No.	Customer's Name (and their Order Number).	Description of Job.	Day Book Folio.
1912 Nov 7	410	John Smith	1,000 Billheads	27
" 7	411	Adams	500 Envelopes pld.	27
" 8	412	Carson &Co	8 pp. Price List.	
" 8	413	Mcae &Co	2 Day Books.	
" 10	414	Geoffrey Carr	100 Visiting Cards.	27
" 10	415	Green &Co	10,000 Balance Sheets.	29
" 11	416	Brown & Brown	Proof 4 pp Circ.	
" 11	417	Henderson	Poster	

Form No. 12.

WORK TICKET or ORDER SHEET.

Work Ticket No. *415.*   Date *Nov 10th 1912*

Customer *Green & C*

Address *Walton Road, London, S.E.*   Day Book Folio. *29 !*

Customer's Order No. *1879*   Estimate No. *278*

	QUANTITY.	DESCRIPTION.	SIZE.
Show 3 Proofs.	*10,000*	*Balance Sheets. L. Post. 4to fly.*	
		L. Primer, printed Black, Folded to endorse & inserted in envelopes stuck down.	
Nov. 11.	*Revise*		
" 13			
" 14	*Press*		
		Deliver 2.500 Nov 17th to Green & C	
		Balance Nov 18th to The Addressing Agency. High Street. E C.	

Delivery wanted by...................................

DATES RECEIVED BY EACH DEPARTMENT.

COMPOSING.	MACHINE.	BINDING.	PAPER.
Nov 10th	*15/11/12*	*Nov 14th*	*15/11/12.*

DELIVERIES.

Date.	Quantity.	Delivery Note No.	Date.	Quantity.	Delivery Note No.
Nov 17.	*2500*	*2409*			
" 18	*7.500*	*2723*			
	10,000				

This Work Ticket may be printed as a single form or as an envelope. The envelope is convenient for holding copy, pattern, etc., and should accompany the job through the works. The particulars should be sufficient for the work to be carried out without further instructions.

Form No. 13.

PRINTING MACHINE UNITS.

Example showing how to find the hourly rate of each machine.

nk is not included on this form, and the value of ink used and the proper percentage for overhead expenses must be added to the job cost sheet to find the cost of each job.

1. Machine	2. Original Cost of Machine	3. No. of Units for each Machine. Equals 1 unit for every £50.	4. Average hours run in the week	5. Fair hour Units per week Col. 3 × Col. 4.	Average weekly Total Cost of Dept. ... 26.15.6 Less Chargeable Wages (Total of Cols. 6 + 8) 8.15.6 £ s. d. 14.0.0	6. Machine Unit Hour Cost (Hourly unit value = .255 by Col 3).	7. Feeders' Wages.	8. Minder and Ap previous Wages	9. Minder's Wages, if an Apprentice	10. Minders Wages with adjustment to include Apprentice	11. Total Weekly Wages paid as each Machine after Adjustment	12. Wages per chargeable hour.	13. Machine hourly rate = Wages + Machine Unit hour Cost
2 D Double	£ 550	11	46	440	Divide this amount by the total of Col. 5, and thus will give Hourly Unit Value = 2.55	2/4	10/6	45/-	45/-	37/6	50/-	1/3	3/7
2. Cr Wharf	325	7	37	259		1/6	10/-	37/6	37/6	31/6	40/6	1/1½	2/7"
" "	325	7	37	259		1/6	10/-	19/-	35/6	29/6	40/6	1/1½	2/7½
Royal "	225	5	40	200		1/1	7/6	35/6	35/6	24/6	22/3	6¾"	1/7"
Demy "	200	4	40	160		10¼	7/6				22/3	6¾	1/5
				1318			£2.7.6	6.8.0	7.13.6	6.8.0			
							£ 8.15.6						

In Col. 10 the minder's wages have been adjusted, so that the advantage of lower wages paid to apprentices is not taken by any one machine or customer. The advantage is spread over the whole room. The chargeable hours of each machine at the hourly rates in Col. 13 should be entered on the Job Cost Sheets and the Value of Production form.

This form should be made out and the hourly rates continued for a year. The rates can then be checked again. Every machine in the department should be included in this form.

This form is filled up with the details of a different business to the previous forms, in order to show how the hourly rates for large and small machines can be found.

Form No. 14.

BINDING MACHINE COSTS.

Example showing how to find the net machine cost per chargeable hour.

Only the net cost of the machines should be included in this table.   To the net machine cost per chargeable hour add the net wages paid per hour, and to this total add a percentage to cover departmental and overhead expenses, and the result is the hourly rate for each machine.

	ONE QUILLOTINE CUTTING.			ONE THREAD SEWING.			TWO WIRE STITCHING.		
ORIGINAL COST OF MACHINES	£100.			£170.			£45.		
FLOOR SPACE	8 x 8 = 64 feet.			4 x 5 = 20 feet.			5 x 10 = 50 feet.		
	£	s.	d.	£	s.	d.	£	s.	d.
COST PER ANNUM.									
Depreciation, 10 %	10	0	0	17	0	0	4	10	0
Insurance, 12/6 %	0	12	6	1	1	3	0	5	8
Interest on Capital, 5 %	5	0	0	8	10	0	2	5	0
Rent, rates, taxes, lighting, heating, &c., @ 2/- per foot	6	8	0	2	0	0	5	0	0
Power	10	0	0	1	0	0	3	0	0
Repairs and Sundries... ... ... (Oil, knife-grinding, parts, needles, &c.)	5	0	0	2	10	0	2	0	0
Total net machine cost per annum, without wages or departmental and overhead expenses ...	52) 37	0	6	32	1	3	2)17 0 8 / 8 10 4		
Do. do. per week ...	0	14	3	0	12	4	0	3	3
Average chargeable hours per week ...	43			33			32		
Net machine cost per chargeable hour ...	4d.			4½d.			1¼d.		

This form should be made out and the hourly rates continued for a year. The rates can then be checked again. All the machines in the department should be included. Several machines of the same type can be included in one column, so as to obtain an average rate. These figures do not refer to the business dealt with in the previous forms.

Form No. 15.

PROFIT AND LOSS ACCOUNT

For twelve months ending September 30th, 1912.

Dr.		£	s.	d.	Cr.		£	s.	d.
Work in Progress (at beginning of year)		250	0	0	Sales		3,000	0	0
MATERIALS—					Work in Progress (at end of year)		100	0	0
Stock on hand at beginning ...	£380 0 0								
Add Purchases during the year ...	670 0 0								
	1,050 0 0								
Less Stock at end	300 0 0								
		750	0	0					
WAGES		1,050	0	0					
EXPENSES—									
Rent, Rates, Light, Heating, &c....	130 0 0								
Power	20 0 0								
Depreciation	120 0 0								
Fire Insurance	10 0 0								
Workmen's Compensation ...	2 10 0								
Repairs	20 0 0								
Departmental Sundries	10 10 0								
Interest	100 0 0								
All other General Expenses—									
Proprietor's Salary £208 0 0									
Clerk 32 10 0									
Stationery, Postage,									
and Telephones 12 0 0									
Carriage 15 8 0									
Discounts ... 11 12 0									
Bad Debts ... 21 10 0									
Spoilage 8 15 0									
Advertising ... 7 0 0									
Legal and Bank									
Charges, &c. ... 5 5 0									
General Trade Ex-									
penses 40 0 0									
	362 0 0								
		775	0	0					
PROFIT		275	0	0					
		£3,100	0	0			£3,100	0	0

(The Printers' Cost-Finding System, 2nd ed., July 1913,
Specimen Forms).

PRINTERS' COSTS
(*Review in* The Accountant, *1913*)

In another column of the present issue we reproduce the recently-issued report and recommendations of the Special Committee of the Federation of Master Printers and Allied Trades of the United Kingdom on "Costs and Charges," which were considered, and very favourably received, at the Congress of that body held at Kingsway Hall on the 18th and 19th ult. This Congress was, we believe, the largest gathering of Master Printers that has yet been held in this country. More than one thousand members were present from all parts of the Kingdom, and as a result of the meetings a general impression was left that the new cost-keeping system would in the end be adopted by the printing trade as a whole. At the same time, it was recognised that its introduction would probably be somewhat gradual, and we understand that, with a view to popularising it, it is proposed to arrange for lectures to be given at the different printing centres, and for the Special Committee of the Federation to appoint an expert who will be prepared to explain the system in relation to the working of individual printing offices, or groups of offices. Further, we are informed a permanent Costing Committee is to be appointed to continue the work. With the same object in view, a further Congress will be held in Scotland during the present month.

In support of the proposed system, it was mentioned by one delegate, who had employed it in connection with his own business, that its working cost amounted to only 1.4 per cent. on the total expended upon wages and materials, while the gross profits were increased by 25 per cent. It is, of course, very obvious that such a startling result as this could only be achieved by increasing selling prices; but we take it that the primary object of the Master Printers' Association is to demonstrate (particularly to the smaller houses) that prices at present charged are frequently far less remunerative than they are believed to be. This we can well understand, for one has only to compare estimates for the same work received from, say, half-a-dozen printing firms selected at haphazard, to realise that their prices cannot be based upon any accurate system of actual costs - or, indeed, upon any uniform system of what costs ought to be - which is the more surprising in that the great bulk of labour costs in the printing trade are paid to union men, and are, therefore, at fixed rates. The explanation presumably is that hitherto Master Printers for the most part have based their costing upon no accurate system depending upon facts; a reflection which suggests that they have not hitherto availed themselves of the services of the accountancy profession nearly so widely as they might with advantage. Even this, however, may be explained by the admission of one of the delegates, that, "in the past, printers have been excellent craftsmen, but not very good business men."

The system of costing has now been under consideration for more than a year and while its practical aspects have been well looked after by members of the Committee who could be relied upon to deal satisfactorily with this side of the matter, the technical aspect of the subject has (we understand) been reviewed by Mr. A.C. Roberts, F.C.A., the auditor of the Federation. Very wisely, as we think, the system has been designed as something quite independent of the financial accounts, and, therefore, is one which can be introduced at any moment irrespective of the precise system of accounting in force and without of necessity requiring any remodelling of that system; but it may, perhaps, be questioned whether it is altogether wise so utterly to divorce the accounting and the cost departments of a business, for to do so involves at least the danger that each may continue to work absolutely independently, and thus to build up different, and wholly inconsistent, results. Naturally, only confusion can result, if the costing department shows a satisfactory rate of profit on each job undertaken, while the accounts department continuously show an unsatisfactory position of affairs. Yet, as our readers will doubtless confirm, this is not altogether infrequent effect of the introduction of a costing system in cases where no effort has been made to connect the costing with the accounting records.

We recognise, of course, that the mode of connection must of necessity, vary according to the nature of the accounting system, and that, therefore, the task of introducing a uniform costing system is far simpler if no such connection be attempted; but it should, we think, be pointed out that, to produce satisfactory results, this connection must not merely be attempted, but also established. It is for the auditors employed by each printing house to arrange with their respective clients how this may best be accomplished, having regard to their particular circumstances.

Speaking quite generally - and we do not profess to discuss the system recommended in detail, inasmuch as it sufficiently explains itself - we are glad to observe that the Association of Master Printers has not been content with something altogether too rough and ready to be really satisfactory. The importance of dealing more or less in detail with Oncost (or, as the Association calls it, "Overhead Expenses") is fully recognised, and although the system described, may, perhaps, cause some misgivings to those Master Printers who have no aptitude, and little liking, for accounts, the fact that the system can be worked satisfactorily at a cost of less than 1 1/2 per cent. on the total paid out on wages and materials should suffice to convince even them that, whatever it may do or not do, it is not likely to involve them in any serious expense on clerical labour. After all, all necessary clerical records in connection with a very adequate costing system can be undertaken in many cases without making any addition whatever to the clerical staff. But, however that may be, there are other ways of making money than by increasing one's turnover and quite one of the most efficient of these is by reducing operating expenses. Moreover, the last-named has this somewhat important advantage over the first-named, that it does not involve any increase in working capital to operate it, but rather the reverse.

At the same time we are not at all sure that the term "costing" in connection with the proposed scheme is not a complete misnomer. Properly applied, the term "costing" relates to a system of apportioning actual charges to find out the proportion thereof that ought properly to be allocated to each piece of business undertaken, with the objects *inter alia* of seeing what kinds of business contribute most generously to the actual profits earned, whether the supervision of the work done has been such as to produce the best results obtainable, and whether the original estimates of cost are all that they should have been. The system now before us is, as we understand it, rather a system of estimating the cost of proposed work rather than of determining the actual cost of the work performed. As the first step in the direction of more enlightened and up-to-date methods, it is no doubt a most desirable one, and one which may be relied upon to prove its value in practice; but at the same time it is, we think, distinctly in the nature of a half-measure. There is little to be gained by knowing what certain work ought to be done for, if at the same time one has no means of knowing what it cost to do the work when obtained; and it is here, we think, that in practice a discrepancy between expected results and experienced results may be found in some cases to produce disillusionment and disappointment. There would be no harm in this, could one reasonably expect that the next step would be to complete the system by establishing an equally reliable record of achievements, to compare with the record of expectations; but unless the incompleteness of the system at present advocated be quite clearly explained it is, we think, much to be feared that in many cases those who experience this discrepancy in practical working will be inclined to abandon the whole process as useless, instead of turning their attention to completing it. That such discrepancies must necessarily arise in practice will, we think, be very apparent to our readers, for they are inherent in the very nature of the system; but it is hardly to be expected that they will be equally obvious to the average Master Printer - nor, indeed, would it be fair to expect such a standard of knowledge on his part concerning what is obviously a highly technical subject. While, therefore, we wish the movement every possible success, and while we heartily congratulate all those responsible for carrying it so far forward, we should like to impress upon them that it is only due to themselves that they should without further delay explain that, so far, they have not formulated a costing system at all; but only a system for estimating costs on a uniform basis. Probably they are quite right in regarding the organising of printing estimates as the first, and most important, step in the right direction; but they would be quite wrong, if they were to suppose for one minute that, this accomplished, nothing further remains for them to do in order to place matters upon a satisfactory basis.

(*The Accountant*, 8 March 1913, pp. 329-332).

*

PRINTERS' COSTS
(*To the Editor of* The Accountant, *1913*)

Sir, - I have read the leading article on "Printers' Costs" in your issue of the 8th inst. with great interest, and, as a member of the Costing Committee, wish to thank you for the prominence you have given to the subject, but with some of the statements made, and conclusions arrived at by the writer of your article, I am quite unable to agree, and feel that if they go uncontradicted they may create prejudice against the system.

Firstly, the accounting and costs departments of a business are not in this system "utterly divorced." Form 1 is compiled from the Profit and Loss Accounts of the previous year, or average of years, "except where the actual expenses are known to have changed. I can say from experience that if the system is properly worked the monthly and quarterly profits of the business can easily be ascertained, the total of which will be found, when the annual Balance Sheet is got out, to approximate very closely to the profit there shown for the year. I regard this as of great value in keeping a grip on one's business, and I fail to see on what grounds the writer of your article suggests that this costing system may build up a different result from the accountancy department of the business.

Secondly, I venture to think that the statements made in the right-hand column on page 331 show that your leader-writer has not made himself acquainted with the system. The object of the system is to enable the printer to correctly ascertain the complete cost of each hour of labour, or piece unit he sells, and to know that unless there has been any considerable alteration in the conditions of his business anything obtained beyond such cost is profit, and that the total of these profits for any period is subject only to such adjustment as may be necessary in connection with the departmental surpluses or deficits shown by Form 5 for a similar period. By the system it is only a matter of classification to ascertain "what kinds of business (or what selling channels) contribute most generously to the actual profits earned," and by the weekly Statement of Cost of Production, Form 3, the management can see at a glance whether the supervision of production has been satisfactory.

Estimates being given on the ascertained hourly or piece unit rates will only vary from the costs returned on the job if the estimator has been at fault in calculating the time necessary on the job. The management will naturally watch closely the costs returned on estimated jobs, and compare them with the estimates.

I may add that the company with which I am connected has used a similar system for the last five years, and it has been found in every respect satisfactory.

I am, your obedient servant,
WILLIAM A. WATERLOW.

London, 17th March 1913.

(To the Editor of The Accountant)

SIR,- Mr. R. J. Lake, Secretary of the Federation of Master Printers and Allied Trades of the United Kingdom, has drawn my attention to the leading article in your issue dated the 8th inst., with reference to the "Costing System" recommended by the Special Committee at the Congress of Master Printers held at Kingsway Hall last month.

Before attempting to criticise or to damn with faint praise the costing system referred to, it is surely not unreasonable to have expected the writer of your article to have made some effort to study the subject, and, if not clear on any of the points involved, to have referred either to the Committee or myself for information.

The "system" adopted by the Committee was decided on after full consideration as being the most simple and most effective for the printing trade. In many respects the system is identical with that now in use in America, and it is one that I have introduced into many printing houses with the very best results during the past ten years.

The writer of your article states that the system is quite independent of the financial accounts.

If present at the Congress he evidently did not follow Mr. Howard Hazell's masterly speech, and he evidently has not studied Form 1, which explains the method of arriving at departmental assessments entirely from the figures of the Trading and Profit and Loss Accounts.

As a matter of fact, the Costing System is built up out of the actual figures shown by the accounts of a business, and is adjusted from time to time in accordance therewith.

Your leader-writer suggests that the system the Committee recommended is an Estimating and not a Costing System, and that it *"does not determine the actual cost of work performed."*

The whole system is designed to enable the printer to ascertain the cost per hour of every man and every machine in all departments.

These costs or hourly rates are checked weekly with actual wages paid, plus the weekly proportion of departmental charges and overhead expenses.

One of the most important points of the system is that it determines "actual" costs as distinguished from "estimated" costs, as every hour spent on a job must be charged thereto.

Another point is that at any time of the year it is quite easy to ascertain the amount of wages and general expenses recovered by charges to jobs.

I believe the system recommended to be "foolproof" as far as any cost system can be, and it can easily be adapted to the many variations and difficulties of the printing trade.

As long as the Purchases Day Book, Petty Cash and Wages Books have columns for analysis, the ordinary books of account are all that are actually necessary for the average printer to use.

I enclose copy of Job Cost Docket, Form 12, and also Forms 9, 10, and 11, showing the value of production in the departments dealt with, which you have not reproduced, and which may explain some of the points dealt with in your article.

I shall be glad if you will kindly publish this letter, as, although many printers may not read *The Accountant*, their attention may be drawn to your remarks, and I feel sure it is not your desire to belittle the system advocated without giving it proper consideration.

Yours faithfully,

ARTHUR C. ROBERTS, F.C.A.

London, 17th March 1913.

[Our correspondents are, we think, unduly sensitive. The report submitted to us has been reproduced verbatim in these columns, and, if we have failed to do justice to it, the fact will be patent to all who care to read. Our point is that while the system is all right so far as prime cost is concerned, it substitutes estimates (based on past years) for facts in respect of oncost charges. It is all very well to say "except when the actual expenses are known to have changed," but there is no clear statement as to what is to be done then - and they are sure to have changed. The fact that the system is in use in America does not impress us.-ED. *Acct.*]

(*The Accountant*, 22 March 1913, pp. 435-436).

*

SELLING THE COSTING SOLUTION

The Costing Committee enthusiastically pursued its remit of promulgating the widespread usage of the printers' cost-finding system following the First Cost Congress of February 1913. Within days of the Congress the presidents of local associations of master printers were urged to establish their own costing committees and organize district costing campaigns. The Costing Committee was buoyed up by reports of the experience of American master printers which continued to suggest that the ideal of universal adoption was not chimerical.

The Second Congress in 1914 was informed that a reasonable degree of progress had been made in the British costing movement though this was insufficient to "effect a millennium in the trade". Thereafter a series of situational factors conspired to hinder the twin objects of the complete organisation of master printers and their usage of the prescribed costing solution. The effects on costs and the demand for printed products of World War I and cyclical downturns during the early 1920s and early 1930s, engendered renewed bouts of price cutting.

In response to these impediments to collective behaviour and commonality among printers, the propaganda techniques employed to sell costing became more varied. The Federation's journal, The Members' Circular, *was a principal medium for carrying material designed to alter printers' attitudes towards costing. Reports on the costing movement in trade and professional journals were encouraged in order to enlist the support of other potentially interested parties, such as accountants.*

The Costing Committee and its Propaganda Sub-Committee periodically conducted studies to ascertain the extent to which the Federation Costing System had been installed in printing firms. The generally disappointing results of these surveys by the 1930s caused some to question the wisdom of continuing with the costing campaign. Such criticism resulted in a more education-centred approach to propagating usage of the costing system, and, in the period of post-war prosperity a greater a degree of success was achieved. Fifty years after the First British Cost Congress it was reported that more than one-half of member printing firms were using the Federation Costing System.

COST FINDING

(Letter to the Presidents of Local Associations of Master Printers,
1913)

The Members of the Cost Committee have instructed me to convey to the Trade generally their appreciation of the magnificent response that was accorded to their invitations to attend a Cost Congress. The fact that some twelve hundred were present speaks for itself.

They are now, seeing that such deep interest has been awakened, more than ever anxious to quickly follow up, the suggestions they have laid before the trade, with the object of getting them universally carried into effect.

My Committee, feel most strongly, the necessity of emphasising the great advantages that may be expected to follow, if the trade will only cooperate in this matter. And further they do not think that as favourable an opportunity of acting together is likely to occur again, at least for many years to come.

You are particularly requested as President of your Association to kindly use your influence to induce your members to carry this movement into effective operation.

I venture to enclose some Resolutions, which my Committee trust each Assn. will adopt at an early date.

I shall esteem it a favour if you will please let your Secretary advise me of the steps actually taken by your Association.

> Yours faithfully,
> Secretary.

Resolutions for each Association to submit to its Members

1. That the Cost System as approved at the London Congress be accepted by this Association and that its universal adoption be strongly urged on our members.
2. That a local Cost Committee be formed to carry out the suggestions made at the Congress, including the fixing of local rates for the various operations.
3. That immediate application be made to the Secretary of the Federation for a Lecture by an expert.
4. That a subscription be specially raised and forwarded to the Federation Authorities as a contribution towards the Expenses

that are being incurred in this matter.

5.   That a Sub-Committee be found to wait upon those local Houses who are not members of this Association to invite them to cooperate in this important movement.

6.   That the thanks of this Association be given to the Costing Committee and that its members be assured of our support and cooperation.

(Costing Committee, Minutes, 26 February 1913)

*

THE SECOND COST CONGRESS
(1914)

AN OUTLINE AND SOME IMPRESSIONS OF A SUCCESSFUL FUNCTION

The master printers of Great Britain have demonstrated, in an unmistakeable manner, their approval of the holding of an annual Cost Congress, by coming to the Second Printers' Cost Congress in numbers equalling the epoch-marking attendances at the First Cost Congress last February, and with their enthusiasm in the Costing Movement unabated. This Second Congress, occupying a day and a half of solid, studious debate, contributes another memorable event to the history of British printerdom, and marks another forward step in the direction of a raising the printing trade to a higher economic position among the industries of the country.

To see the printers trooping in their hundreds into the fine large Kingsway Hall, at 2.30 on the opening day, Tuesday, February 24, was a sight that did one's heart good, and at the close of the Congress, after such a feast of eloquence, wit and wisdom, there could have been none whose estimation of our grand old craft had not been enhanced. An optimistic feeling pervaded the whole proceedings, and the numerous speeches reverberated with a cheerful and unwavering confidence in the rightness of the course printers have taken in placing their business on a sound and scientific basis.

Previous to the opening of the Congress, the speakers and the Costing Committee and also the London members who had volunteered to act as stewards, were entertained at lunch in the Connaught Rooms, adjoining the hall, by Mr. W. A. Waterlow (Chairman of the London

Association). The Chairman on the first day, Mr. George E. Stembridge (President of the Federation), set an admirable example of brevity and lucidity, and gave the key-note to the Congress when he said, "we are business men met for business purposes, to consider business details in a business frame of mind," and that was the spirit that characterised the proceedings throughout. He conducted the meeting with consummate skill, as did also Mr. J. E. T. Allen (Chairman of the Costing Committee), who presided on the second day. Then Mr. Stembridge struck a cheerful note that was re-echoed again and again. "I am very glad indeed," he said, "that we have discovered the great advantages of co-operation, of conference, and of comparing notes one with another on those important matters that occupy our everyday attention." His tribute to the excellent work of the Costing Committee was generous indeed, but is more than justified, as was audibly confirmed by the large assembly. The world-wide interest in the scientifically accurate system they have evolved is evidenced by enquiries from nearly every part of the world, and from other industries than the printing industry.

The first speaker called upon by the Chairman was an Amsterdam printer, Mr. J. H. Binger, Secretary of the Costing Committee of the Netherlands Federation of Master Printers, and a member of the Council of that body. He explained that his Federation being only of recent date, their Costing Committee had not achieved anything yet, but were still studying the question of costing, the economic position of the Dutch printer needing to be raised far beyond its present level, which he described as below zero, a naive remark that evoked the first of many hearty outbursts of laughter. Mr. Binger's impression is that competition is even keener in his country than ours, and he raised another hearty laugh when he explained that "Here people ask for quality first; in my country a great many people ask for prices first and when they've got the lowest price they ask for quality as well." As elsewhere, wages are going up in Holland and prices are suffering from severe competition, and Mr. Binger looks for correction of the latter bad condition in the establishment of a scientific and reliable system, based on similar lines to that of the British Costing Committee. He has used a similar costing system in his own business for ten years to his entire satisfaction.

Mr. J. E. T. Allen's survey of the Costing Movement since its inception was most interesting and inspiring, and he may rest assured that all progressive printers are perfectly convinced that the Costing Committee have both justified its existence and proved itself worthy of the confidence placed in them. A fine spirit of unselfishness and devotion to the betterment of the trade has actuated them all. It must have been some recompense to them to know that - to use the words of Mr. Allen - "Never before has any matter so interested the trade as did the Costing System which we gave you last year; never before did printers congregate together in such numbers to explain that system and hear it explained, and that interest has been well maintained, as is proved by the attendance here to-day." He spoke of the numerous meetings that have been held in different parts of the country - north, east, south and west - how it had brought together printers who hitherto had hardly been on speaking terms, how it had brought recruits to the Federation and had strengthened it by the addition of the formation of new local associations. He spoke regretfully of the mistaken ideas that seemed to prevail amongst some of the work-people, and assured them that the Costing Committee had no idea of introducing anything in the way of a spy on the work they were doing. All they wanted to know was the cost of their jobs, and if they wanted to spy on anybody it was their customers. Although the Federation is growing, he drew attention to its comparative smallness to the whole of the printing trade, and intimated that they had a great many good movements ahead, the degree of success in grappling with which was dependent on the strength of the Federation. Then came an eloquent appeal for financial support of the movement. With seven or eight thousand printers in the United Kingdom, and 2,000 members in the Federation, 400 subscribers to the Costing Campaign does looks a small proportion.

"The need for a universal adoption of a Costing System and what it will do for the Trade" was the substance of a very earnest address by Col. Wright Bemrose (chairman of the First Cost Congress and past president of the Federation), who described the welfare of the Trade as wrapped up in the movement, in a commendably brief address. The intrinsic importance of the matter was most ably put and convincingly argued, and the evils which have for so long burdened the printing trade and, alas, still continue too frequently to prevent the attainment of its rightful position among the more prosperous industries of the country, concisely reviewed. No one will dispute his assertion that the enormously varying differences in prices is the worst evil with which we, as a trade, have to contend, nor the claim that a system of costing in such a complicated business must be absolutely accurate and thorough. Then, again, it is evident, if the evils of the trade are to be overcome, there must be one universal system throughout the trade. A system based on the fruit of the best thought and experience of the two great English-speaking countries - England and America - adapted to suit the requirements of this country, and made elastic enough to fit any sized business, having been evolved and now being widely utilized, it only remains for it to be standardized and, above all - this point Col. Bemrose strongly emphasized - for printers to stand by the results they obtain by it. He pleaded for a discarding of that reprehensible practice, the fill-up order, for more hearty co-operation, more loyalty and greater respect for each other.

The Federation Cost System was then briefly outlined by Mr. A. E. Goodwin (secretary and organiser to the Costing Committee), who accentuated to an obviously interested and appreciative audience its simplicity and elasticity. There had been, he said, no necessity whatever to depart from the principles laid down at last year's Congress, but, while last year the system was demonstrated from that platform, to-day it was in operation all over the country. He had seen it at work amid the Royal surroundings of Kings Lynn, in quiet cathedral cities, in staid and solid agricultural districts, and in the busy hives of industry in Lancashire. It was in use in an office where there were three men and a boy, and also in a plant employing about 1,000 hands. There was no printing business under the sun to which the system could not be applied. Regarding the daily dockets, a prominent trade union leader had stated on this matter that he had the gravest possible objection to his labour being sold below cost, and no doubt every self-respecting worker held the same objection. This was the spirit they wanted to see in all the men, but it was only by means of this costing system that they could ensure their labour not being sold below its cost. (Applause.)

Mr. W. Howard Hazell, fresh from his visit to the States, and with vivid impressions of the New Orleans Convention of the United Typothetae of America, received quite an ovation, and gave an admirable and inspiring account of what American printers have done in costing, and the benefits they have derived. He found "about 700 jovial printers," and he spoke of the great hospitality accorded him. In the States their efforts with regard to cost keeping and the "almighty hustle" put on, were known and admired as a great work. (Laughter and applause.) In the States they were very well satisfied with the work that had been accomplished by the costing system. 3,000 printers had installed the system as a result of the campaign of the last five years, and it was more than the total membership of the Typothetae, and as a result prices had been steadied. Competition had not ceased to exist, but was now on sane and reasonable lines. A leading printer in Chicago told him he used to consider his cost of composition was about 3s. 3d. per hour, but now he knew it cost him 6s. This was followed by a lucid and detailed exposition of a model Estimate Form, which includes one or two features of the American Forms, but adapted to suit English printers. A good point was made in this capital address when Mr. Hazell stated that, in America, one printer in ten has installed the cost-keeping system; in the United Kingdom, after one year, one printer in seventeen has adopted the Costing System. He asked "Is not that a record of which we may well be proud?" the answer being an outburst of loud applause. Mr. Hazell showed all his unrivalled skill in answering questions, and these were quite numerous during the Congress, a healthy sign that has characterised the movement from the start.

When the plaudits which followed Mr. Hazell's speech had ceased, the chairman read out a verse which had been passed up to him parodying "The British Grenadiers" ("Some talk of Alexander and some of Hercules," etc.).

> Some talk of Senefelder and bills of laundres-sees,
> Of Caxton and of Franklin and such great names as these;
> But of all these ancient heroes not one could preach so well
> The hour-rate, hour-rate, hour-rate, as the British Seer Ha-zell.

. . . Mr. Bernard J. Biggs (Darlington) and Mr. H. Harland (Hull), the one a small printer employing twenty-five hands, and the other a larger printer employing about two hundred hands, gave encouraging accounts of what the Federation Cost System does, as witnessed in their individual cases, both proving enthusiastic advocates of it. Mr. Gronow (Waterlow Bros.) characterised estimating as the twin sister of the Costing System, and, as a practical printer, considered it a most important factor in these competitive days, but he was insistent on the estimator being a practical printer as distinct from the merely theoretical estimator.

Then the Federation Cost System as applied to a newspaper house was very instructively explained by Mr. H. Wirren (Doncaster), who claimed that if there was one office more than another which ought to install the Cost System it was that which combined a weekly newspaper with a general jobbing business. He was followed by a smaller printer, Mr. G. H. Pindar (Scarborough), employing from fifteen to twenty hands, who told, in an excellent speech which was loudly applauded, what the Federation System had told him, and how he had listened and profited by the information. Recalling King George's well remembered saying that the printer was the "friend of all" he asked why the printer was not his own friend - a remark that went home. Chargeable Time was in the hands of Mr. S. A. Penny (Southampton), who clinched a good speech with an anecdote that brought down a burst of applause: a father's advice to his son, "My son, consider the postage stamp; its ability consists in sticking to one thing until it gets there."

After the luncheon interval (during which 120 to 130 of the company were the guests at lunch of the three secretaries: Mr. Lake, Mr. Thomlinson and Mr. Goodwin), Mr. R. A. Austen-Leigh (Spottiswoode & Co.) gave one of the most brilliant addresses of the Congress on the future of the Costing Movement and the need for ungrudging support and therewith of the financial requirements. The burden of his message, briefly, was that though 500 master printers had availed themselves of this system, obviously 500 printers with a costing system cannot effect a millennium in the trade. Then what can be done to make more printers take the matter up ? There are a great many quite small printers who still think the system too complicated; hence an attempt must be made to evolve some plan that will benefit the one-man shop. Developing his argument he advocated first arriving at standard costs, and then at standard rates of profit to add to that, and finally standard prices to our customers. He threw out many valuable suggestions as to how this could be attained, it being a mere question of co-operation and loyalty on the part of the bulk of printers. Standard customs was another topic suggestively discussed in this thoughtful address. All printers should study the address and also those of Mr. Hazell and Mr. Goodwin in the full report.

REPORT OF THE COSTING COMMITTEE TO THE CONGRESS.
(Condensed)

In submitting a report of the work of the Costing Committee since the Congress held in February last, the first words must be those of thanks for the generous response to the appeal made at the close of last year's meeting. Time and money have been given by printers in every part of the country to further the movement which had such a splendid send-off last year.

It is quite impossible to enumerate those who have assisted in the campaign throughout the country, but the committee fully recognise how heartily they have been supported, and wish to express their thanks to all who have helped in any way.

Immediately following the Congress in London, arrangements were made for district Costs Congresses in Edinburgh, Newcastle, Nottingham, Manchester, and subsequently at Dublin, Portsmouth, Bristol, Leeds and Stockton. All these Congresses were remarkable for the interest and enthusiasm shown, and the number of enquiries which the speakers had to deal with. Resolutions expressing approval of the Federation System, and urging its general adoption, were carried unanimously at all of them.

In April last the committee decided to appoint an organising secretary to carry on the work and continue the educational movement in all parts of the country. They were fortunate in being able to secure the services of Mr. A. E. Goodwin, who had installed and worked a cost system in a large printing and publishing establishment, and who had also a wide experience in accountancy and a special training in methods of costing.

The committee were therefore able to respond to the numerous requests to arrange special meetings to discuss costing problems with local Master Printers' Associations in all parts of the country, and meetings have been held in Swansea, Peterborough, Brighton, Stoke, Leek, Norwich, Bradford, Scarborough, Wisbech, Boston, Weymouth, March, Aberystwyth, Aberdeen, Edinburgh, Glasgow, Dundee, Stirling, Bolton, Hull, Hyde, St. Albans, Watford, Sheffield, Barrow, York, Burnley, Huddersfield, Gloucester and King's Lynn.

A special Costs Conference was held at Liverpool on the occasion of the Annual Meeting of the Federation.

It will be seen that at practically every important printing centre meetings of master printers have been held, and the subject has been well discussed. There are still, of course, a large number of towns it has not yet been found practicable to visit, but, wherever local printers have undertaken to make the local arrangements, the committee have provided speakers, and they hope to be able to comply with all requests for assistance of this nature, and gladly welcome such applications. It has been quite a general experience that, even where only one or two printers take an interest in this movement, it is possible to secure a good attendance, and beneficial results have invariably followed such meetings.

Early in the campaign, the committee realised the importance of securing the hearty sympathy and support of the managers and overseers, and with the ready assistance of the officials of the Printers' Managers and Overseers' Association meetings were arranged with their members in London, Liverpool, Manchester, Nottingham and Leeds, to which all overseers in the respective districts were invited. Many misapprehensions as to the real purpose of the movement were removed by these meetings, and everywhere warm approval of the committee's recommendations was expressed.

Another important step was taken in approaching the City and Guilds of London Institute, with a view to securing that the Federation Cost System should be taught in the typography classes of all technical institutes, as the majority follow the Syllabus of the Institute and prepare students for the Institute Examination.

As the Syllabus was at the time being revised, the Council included a synopsis of the Federation System under the heading of General Management, and senior students will be required to study the principles of costing.

A Conference with the Technical Instructors was held in London, when the System was explained and methods of teaching it discussed.

The Organising Secretary has given lectures at the St. Bride Institute, Camberwell School of Art, the Aldenham Institute, Borough Polytechnic, and Regent Street Polytechnic, and the head Instructors are all greatly interested in the movement. A Conference with the Technical Instructors in Manchester was also well attended.

From the outset the committee have endeavoured to secure the hearty co-operation of the Accountancy profession, and have to express their thanks to many who have given freely of their time to explain the system and methods recommended. Mr. A. C. Roberts, F.C.A., spoke at St. Albans and Edinburgh, and a lecture was given at Sheffield by Mr. A. H. Muir, C.A., who also delivered a series of lectures in Belfast; Mr. W. B. Phillips, C.A., lectured to the Chartered Accountants' Students Societies in Manchester and Liverpool; and Mr. Nickson, C.A., at Leicester, has rendered valuable help.

Mr. W. Howard Hazell addressed a meeting of the London Chartered Accountants' Students Society, and Mr. Goodwin has arranged to address the Incorporated Accountants' Students Society.

In Edinburgh Mr. F. H. Bisset delivered a lecture before the Chartered Accountants' Students Society; and Mr. J. Hamilton Gray, C.A., and Mr. J. Lauder, C.A., have been appointed Secretaries of local

Cost Committees at Edinburgh and Glasgow respectively.

The issue of the explanatory booklet, with forms suitable for a small business, written by Mr. W. Howard Hazell, with an Introduction by Mr. R. A. Austen-Leigh, gave considerable impetus to the movement, and the demand for it was remarkable. Seven thousand copies were issued, and so clearly and fully is the System explained that many, aided by the book, have been able to install it without difficulty.

An important development was the following up by a personal canvass of several cities and districts where meetings had been held, and the Organising Secretary was thereby able to ascertain to what extent practical effect had been given to the resolutions adopted at the numerous conferences. The results are distinctly encouraging. Out of 225 offices visited, in 12 of the principal centres, 111 are working or intending to adopt the Federation system.

These figures show that in districts where conditions vary considerably satisfactory progress is being made. It must not be assumed that these figures give the total number of firms working the system at present in these districts. It was impossible to call at every office in the large centres, and in many offices the installation of the system has not been completed. At Edinburgh, in reply to a circular enquiring if firms intended adopting the system, 31 out of 50 replied in the affirmative. In Belfast a similar enquiry was made, and the larger firms reported that they were taking action to adopt the Federation recommendations. It is certain that over 500 firms have taken some action.

During the ensuing year, no doubt, steps will be taken to ascertain what firms are working on Federation lines, as the idea of finding out the standard rates for various processes has commanded itself to many. In a Gloucestershire district, average hourly rates of cost have been arrived at, and in one Lancashire town, and also a Yorkshire district; in London, enquiries are now being made.

Numerous enquiries having been received for assistance and special advice in the installation of the system, the committee approved of a scale of fees which would enable the proprietor of a small plant to obtain the services of the organising secretary at a very moderate figure.

As regards finance, it will be seen from the accompanying statement of receipts and expenditure that a sum of £1,484 16s. 6d. was received from subscribers and guarantors, the full amount of the sums guaranteed having been called up and paid. The total number of guarantors and subscribers was 413. Liabilities have been incurred in connection with the second congress, etc., since the accounts were prepared, which will practically dispose of the balance in hand. It is obvious that, if the campaign is to be conducted with the same thoroughness and vigour, it will be necessary to raise at least £1,500 this year and, to secure this amount, many new subscribers will be required to supplement the efforts of those who have hitherto so generously supported the work.

World-wide interest is being taken in this movement. Enquiries for literature have been received from Australia, America, New Zealand, South Africa, India, Egypt, Holland, Germany and Sweden. Mr. R. A. Austen-Leigh was present at the Swedish Costs Congress, and Mr. W. Howard Hazell at the American Printers' Costs Congress at New Orleans. Holland will be represented at this second congress.

In concluding this report, the committee venture to say that they trust that the measure of success, which has attended their efforts, warrants them appealing to all, who are interested in this attempt to place the industry on a sounder basis, to continue their support, and that, if this is done, at the close of next year the progress made will be even more marked.

(Members' Circular, February 1914, pp. 64-72).

*

PRICE-CUTTING AND THE WAR
(1915)

The world is at war, and printers, like most other manufacturers, are suffering severely. But serious as the results of the world-war may be to the master printers of this country, there is another and an even more dangerous form of war - the unreasonable internecine war of price-cutting. Keen as the competition has been in our trade in the past, it has probably never reached such a level as in the last few months, when printing prices have moved in inverse ratio to the cost of production, and buyers of printing have set one printer against another, until the scramble for the work has resulted in an insensate and undignified Dutch auction. Before the war it was generally recognised, and printers' balance sheets proved the fact, that the printing trade was not a highly profitable industry, and, since the war, the cumulative effects of a decrease in the quantity of work, increases in the cost of materials and production, and in many cases a reduction in prices, must have reduced former good profits to zero, and converted small profits to very serious losses.

If these calamitous results were the direct and inevitable consequences of the war, the average British printer would not complain, but would hope for a speedy peace and better trade. But the position is largely due to the eager scramble for work, at any price, regardless of cost, and a lack of that necessary co-operation and cohesion amongst printers which would end or mend most of the ills from which our trade suffers. Our products are not luxuries, but principally necessaries, and we have little fear at any time of foreign competition, and none at all at present; and yet, in spite of these advantages, we are finding that our prices are being undercut and our profits reduced. In many other industries with which we deal, there is greater firmness and determination to maintain a reasonable price. Our paper, coal, oil and many other materials have been increased in price, and these increases have probably not yet reached their limit.

The danger of this price-cutting during the war is not limited to the war period, but will probably last long after peace has been signed, and will hinder much of the good work that has been achieved by the Costing Campaign. If our customers receive a low quotation now, their prices and their methods will be based upon it, and they will, in the future, strive to obtain similar work at the same or lower prices, and hunt for estimates until they find a printer willing to take the work at war prices. In some trades a reduction in price greatly increases the demand for the products of that industry, and owing to the greatly increased demand normal profits are maintained, but this is not the case in the printing trade, as a reduction in charges - particularly under war conditions probably does not increase the demand at all. There is a certain amount of printing required under the present restricted conditions of trade in the country, which would be done (if prices were maintained) at their normal level, but, owing to price-cutting, the same amount of work is being done far below cost, to the permanent injury of the printing trade, and to the immediate benefit of the customer, who profits by this wilful price-cutting.

The contemptuous opinion of printers and their prices, which is held by many buyers, is shown by the remark of a customer who was striving to reduce the printer's price, who said: "You printers always have three Prices. Your first price is the one you send in, the second price is the one you will take if we offer you the job below your price, and the third price you will ultimately accept if we say we are going to take the work away." As long as this and similar opinions are held by those with whom we deal, the printing trade will suffer, and neither make the profits nor attain that commercial position which we ought to hold.

The remedy for this great and growing evil lies in our own hands. All the world knows that the price of commodities from coal to codfish, and from wool to wood, has grown greatly, and if material enters largely into a printer's products, his customers would, in many cases (if the facts were explained) be willing to pay the increased cost of materials. It is a definite increase, which is easily comprehended by a layman who cannot understand the meaning and high cost of a compositor-hour. The surest way, however, of maintaining reasonable prices, is for all printers to give a fair price, and stick to it; to respect another printer's imprint, and to co-operate with his fellow master printers in this most urgent matter. A few selfish and short-sighted printers, by quoting reckless prices for far more work than they will ever obtain, may do incalculable harm, by lowering prices all round; but if every printer would realize the necessity for greater firmness in quoting prices, and greater co-operation with others, we should pass through this national crisis with far less injury than seems likely to be the case.

(*Members' Circular*, March 1915, pp. 119-120).

*

STOP THE LEAK!

You know there is one because stock does not agree with records. Perhaps you have no record — then you need to get busy. A form for checking issues of stock can now be obtained, as well as all costing forms.

Cost & Charges Committee, 24, Holborn, E.C.1.

(*Members' Circular*, November 1920, p. 302).

A Memento of the Fourth Cost Congress

(*Members' Circular*, July 1916, p. 176).

The United Typothetæ of America.

HOW TO BE SAVED FROM FINANCIAL SHIPWRECK.

(*Members' Circular*, April 1918, p. 122).

A TIME OF CRISIS
(1921)

The world-wide trade slump is being felt very keenly throughout the Printing Industry, and as pointed out in a recent issue of the Circular there is a growing tendency on the part of many printers to revert to the old tactics of price-cutting in an endeavour to entice orders into their works.

There is no doubt on this point, for the members of the Costing Staff come across new evidence every day, in every locality visited. Orders are being taken at less than cost.

What we must do now is to face facts, see, if we can, what effect this "cutting" will have upon the industry, and endeavour to find a remedy.

Two main points stand out for consideration.

(1) There are not, for the moment, enough orders to go round.

(2) There is a falling market in paper and materials.

Either of these two conditions make the position of the printer rather difficult, but the two combined make the position critical.

We need not consider the cause of these conditions, as most people are familiar with that side of the question. What we are concerned with is the effect.

The falling off in orders has the tendency to make costs go, up. The hourly rates we are now working on have been established through a period of great activity, and it should be obvious to anyone that they will be found too low, when staffs are reduced or short time is worked. Therefore it is quite certain that those who maintain the present rates are due to stand a certain amount of financial loss in the departments until trade revives and normal output is resumed. The printer who reduces his prices now is making things even worse for himself and for the trade in general, for *cut-prices do not create new orders for printing*. The public may be induced to purchase bargains in many commodities for which they have no immediate need, but buyers of print usually confine themselves to their actual requirements.

The present slump being general, it is obvious that he who endeavours to fill his office with orders at reduced rates is bound to take work away from firms who need it as bad, if not worse, than he does himself. This can only result in retaliation, and though a few orders may be stolen by these bad tactics, he will find that in the long run he is just about as slack as he would have been, with the added irritation of a loss on every job.

Now let us look at item number two, the falling paper market. Almost every printer in the country is heavily stocked with paper, for the most part purchased at prices very much higher than to-day's quotations. This means a direct financial loss to a great many firms, and is all the more reason why the rates for labour should be maintained at a profitable figure. Throughout the war and since, the losses on labour have been more than balanced by the profits on paper. Now we are due to lose these profits and stand a loss in their place.

A price war at the present time will be more disastrous than any we have ever seen in the past.

If ever there was a time when printers should band together for mutual protection and confidence, now is that time. It is true that meetings for the past two years have been frequent, almost too frequent, considering that the principal business has been advances in wages. But it behoves every local body to start a vigorous campaign to eliminate as far as possible this disastrous vice of price cutting. Imprint resolutions should be revived and rigidly lived up to. Wherever possible central estimating bureaux should be established to deal with new estimates and scrutinise work that is changing hands. The clause in the Alliance relating to Standard Hourly Rates where the Standard Cost System is not in use should be enforced, and the campaign for more cost systems should be pushed with more vigour than ever before.

No one can predict how long the present slump is going to last. We all hope it will be of short duration, but even now the industry is face to face with a crisis, and we stand to lose in a few months what it has taken years to gain.-M.H.

(Members' Circular, January 1921, pp. 43-44).

*

AN URGENT MESSAGE FROM THE PRESIDENT.
(1921)

Costing meetings have been held in various parts of the country, and able addresses given by some of the experts of our trade. The scramble for what printing orders there are makes it imperative that the prices quoted should be very accurate, and this result cannot be obtained unless the estimates and the cost are based on a correct costing system. Experience has proved that the best is the Federation

System. Proof of this is forthcoming from the many firms who have installed it. Those, who have not will do well to forthwith get in touch with the Secretary of the Costing Committee, at 24, Holborn, and obtain details.

It is cheap and simple to operate, it is easy to install, it ensures accuracy, it leads to uniformity, it is an ease to the mind and ensures fairness between the employee, the employer, and the customer.

I do appeal to every printer to seriously consider these points, and in this most important direction to put his house in order.

(*Members' Circular*, March 1921, p. 75).

*

A period of unprecedented depression is being experienced throughout the whole country, and every thoughtful individual must realise there is a **DANGER** facing any business which does not possess a precise and accurate knowledge of what it is costing to produce the goods it is manufacturing. **TO THE** Printing Trade, with its many and intricate details, it is of special importance that the cost of each and every operation should be known, and the **TRADE** safeguarded by recognising that greater unity and co-operation between its members will be best secured by adopting a Uniform Standard Hour Rate.

(*Members' Circular*, October 1921, p. 362).

PROPAGANDA SUB-COMMITTEE
(*1924-5*)

Mr. Storey presented the report of the Propaganda Sub-Committee:- A questionaire had been sent out to ascertain how many local Costing Committees were in existence, how many of those on the committees were using the F.C.S., what work was being done and had been done during the past twelve months, how many members were using the F.C.S. and the Standard Hourly Rates. The returns had not been satisfactory and showed only 16% of the membership using the F.C.S.

The Sub-Committee had offered assistance of a kind that would be the most useful to every area. Some had replied that nothing could be done because trade was bad. Two or three Alliances had not sent out the original questionaire of the Costing Committee and the Sub-Committee was endeavouring to get them to do so. They hoped in the course of a few months to have an active campaign throughout the whole country.

(Costing Committee, Minutes, January 1924).

*

Propaganda Sub-Committee. Mr. Storey, reporting on the work of this Committee, presented a summary of the Reports from the various Alliances. These Reports disclosed that in some of the Alliances the percentage of members using the Federation System was 34%, and it fell in other cases to as low as 9%. Mr. Hazell and the Costing Secretary, were of the opinion that where the percentage was low it was due to a great extent to the very large proportion of small firms in these areas. It was agreed to make up a supply of sets of forms for small businesses for circulation at Costing meetings. It was decided to ask that Alliances appointing their representatives to the Costing Committee should be careful to send representatives who were thoroughly familiar with the Federation System. A series of pamphlets and leaflets had been written for propaganda purposes and the order of distribution had been decided upon.

(Costing Committee, Minutes, January 1925).

*

COST REDUCTION
How the Federation Costing System Helps
(*1925*)

USERS of the Federation Costing System have a splendid instrument in their possession which is capable of being used with effect in many directions.

THE FIRST FUNCTION OF THE F.C.S.

Its first function is that of cost-finding. By its means every item of expense is classified and correctly allocated to the department calling for services which have caused the expense so that every operation, whether by hand or machine, bears its proper share of the expense burden. The consumer is thus charged only with an appropriate and equitable share of the cost of production.

Every user of the Federation Costing System who is satisfied to let its usefulness end at this point is like the drowsy householder who in the dead of night hears his faithful watchdog barking furiously but lazily turns over and sleeps.

THE SECOND FUNCTION

The second function of the F.C.S. is that of providing figures for comparative purpose - it first finds the cost and then brings alongside the cost what the producing centres have earned. It is not only capable of doing this for the factory as a whole, but by departments. If the user would find a new interest in business let him carry the matter still further and work the comparison down so finely as to cover the multifarious operations that go to make up a printing order.

These comparative figures when brought together act similarly to the act of "plugging-in" an electric lamp - there is brilliant illumination. What has previously appeared to be a maze of bewildering and cold-blooded statistics now stand out like finger-posts by the roadside pointing in the direction of a certain goal, and in most cases giving the distance that must be traversed before the goal is reached.

THE THIRD FUNCTION

But it is the third function of the F.C.S. which is in all probability its most vital and is certainly its most interesting one-that of cost reduction.

Cost must first be found, then compared and finally analyzed. Just as the chemist breaks down the compound into its several components, so must the manufacturer who wishes to be successful analyse his costs and so refine them that he can prepare a formula that will be a check on the expenditure of the future.

WHO PAYS THE PENALTY?

Competition is fierce and unrelenting, and the stress of the present day is just becoming a struggle for existence in those quarters where there is no respect for a fellow business-man or even for the ordinary rules of the game. The effect of this competition is sadly depleting the finances of many firms and the worst sufferer is usually not the man who loses an order under such competition.

MEETING COMPETITION EFFECTIVELY

The most effective instrument with which to meet competition is that of an efficient Costing System operating its three functions successfully, and not the least important of the three is the last mentioned - cost reduction.

GOOD OUTPUT MUST ACCOMPANY
PROPER RECORDS

The most expensive production unit is not mechanical-it is human. A machine carefully tended will produce good work day in and day out, all things being equal. Temperature may affect the production, but temperament is something that is applied to it. Cost reduction comes most effectively along the lines of increased output, and any amelioration in working conditions that will stimulate a livelier interest in the work under construction will have good results. It has been found, however, that neglect of proper costing records have prevented firms with good production from reaping the full benefit.

BENEFITING BY THE MISTAKES

How many firms examine the kind of errors that are made in the composing-room, as one instance? Will it not be found that the same compositor makes the same kinds of mistakes quite regularly? If by examination such is found to be the case, there is an opportunity to apply a remedy.

COST OF CORRECTIONS

It has been ascertained that as much as ten minutes per 1,000 ens is expended in some houses for house corrections. This is a matter which calls for serious investigation. The expenditure of additional time to correct errors is a state of affairs that will cause serious difficulty in preparing estimates and make great inroads into the profits. Assuming a rate of 2s. 6d. per 1,000 ens for composition, it means an addition of 8d. per 1,000 ens for corrections by hand. It is contended that half this time should be the maximum.

STOPPING A LEAK

The question of make-up is another matter where inefficiency will soon run up costs to an alarming extent, and the provision of suitable making-up "banks" and ample galleys will show good returns.

It is not possible to do more than indicate directions in which economy may be exercised, but a proper use of the Costing System will open up many avenues worth exploration with cost reduction as the goal.

(Members' Circular, April 1925, pp. 116-117).

*

COST FINDING IN THE PRINTING INDUSTRY
(W. Howard Hazell, Past President of the Federation of Master Printers, 1927)

One result of the difficulties of the past few years has been undoubtedly to increase the need for, and stimulate the interest in, correct methods of costing in all manufacturing processes. The great variation in wage rates, the price of materials and transport during the war, and in the post-war boom and slump, have made all previous records of cost of little or of no value. The increased use of labour-saving machinery has also introduced new conditions which affect the cost of production and make the use of proper costing methods still more essential.

In an industry or factory where the goods are uniform, with little variation from a few standard patterns, it is comparatively easy to find the cost of production. When, however, the goods are made only to the customers' order, when every order is different, and the work is produced by men and women, by hand or by complicated labour-saving machinery, the problem of finding accurately the cost of each article is a difficult one. This is the position in the printing industry. In the general printing trade no work can be done until the customer has ordered it, the job must be carried out to the special instructions received, and may vary in value from a few shillings to several thousands of pounds.

About 15 years ago the Federation of Master Printers appointed a Committee to prepare a costing system suitable for the trade. After investigating various methods in use in this country and in America, a system was prepared and finally officially approved as the standard system for the printing trade in this country. The mere recommendation of these methods without some propaganda would have been of little use. Consequently, the Committee selected men with a good knowledge of printing, and trained them as cost accountants, to assist master printers in adopting these improved methods. These officials travel up and down the country and lecture to printers and cost clerks, so that they may fully understand the standard methods. When master printers are willing, the cost accountants will install the Federation costing system in their works, and train the staff in the working of the system.

It is most essential that anyone installing the system should be fully conversant with the technicalities of the printing trade. There are over twenty trade unions in the industry, each with their rules and customs, which vary considerably, and the methods adopted must conform to the agreements arrived at between the Federation and the Unions. The methods of payment may be by time or by piece, and the person installing the system must know what dockets and wage sheets can be used to give the necessary information, and yet comply with the various agreements. In several sections of the industry work is produced by hand, and by complicated machines, e.g. type-setting by hand or by various kinds of machines, each requiring special treatment in costing.

The result of fourteen years of strenuous work has been the elaboration and perfecting of a standard system, which is suitable for, and has been adopted by, firms employing as few as five persons and as many as five thousand. The terminology of costing, the forms for recording chargeable and non-chargeable time, outputs, wages, &c., the allocation of overhead expenses, and many other complicated questions, have been settled, and the system is recognised in this country and the overseas Dominions as the best method for printers. The advantages of the system are that no change is required in the books of account, that the total cost of each job is known within a few hours of its completion, and that each week an account is produced showing the total cost, including overhead expenses, and the value of work produced in each department.

The objects of costing are twofold. *To find the cost and reduce the cost of production.* This costing system by its records automatically gives outputs of different machines and processes, which can be compared with previous records, and the consumption of material, as well as the loss or waste of material, idle time and other details. Guided by this information, the management can make economies and prevent leakages and losses. If a master printer has no proper costing system he cannot know on which job he is making a profit or a loss. Those master printers who have installed the system have found it to be a most potent aid to increase the efficiency of their works.

A difficulty that often arises, particularly in smaller offices, is that the books of the concern have not been properly kept and audited. The result is that the cost accountant, when installing the system in such an office, has to find out, as best he can, many details, such as plant values, depreciation and groups of various expenses which must be separated before they can be allocated to the different departments. The work of the cost accountant begins where the ordinary work of the accountant and auditor is finished, and his task would be much easier if every printer had his books properly audited each year. The cost accountants always recommend to every printer in this position that he should engage an accountant to audit his accounts, so that the annual revision of the costing figures may be simplified, and agree with the Trading Accounts and Balance Sheet. If they follow this sound advice, the result will be beneficial to them, and, incidentally, to the accountants they engage to do this work.

On the other hand, accountants who audit the books of any printer who has not adopted the Federation costing system would do a service to their clients if they recommended them to install the system, which has stood the test of peace and war, and the even more difficult times of the aftermath of war, and which brings the great advantages of accurate cost-finding and increased efficiency. It may seem strange that such a recommendation should be necessary after fourteen years of active propaganda for improved methods of costing in the printing trade. But conservatism is to be found in industry as well as in politics, and there are a large number of master printers who have not yet availed themselves of the services of the officials appointed by the

Federation to install this complete and satisfactory method.

Printers and other manufacturers are very naturally influenced by the advice that is given to them by their auditors on all questions of accountancy and similar matters. The great need of the printing as well as of other industries is to use more scientific and accurate methods, to increase efficiency and reduce costs. The Federation Costing System does very greatly promote these essential industrial aims, and if all accountants would urge their clients to adopt these improved methods, they would be benefiting the printers and assisting the campaign for higher efficiency and lower costs.

The Costing Secretary of the Federation of Master Printers, 7 Old Bailey, London, E.C. will be glad to give information to any accountant about the system, and the services which can be rendered by the cost accountants of the Federation.

(*The Accountant*, 10 December 1927, pp. 783-784).

*

(*Members' Circular*, August 1930, p. 331).

PRICE CUTTING
(*1930*)

From all districts reports are coming to hand of serious price-cutting. One well-known and old-established firm state that it has never been worse in their recollection. Reference is made to the matter in various trade journals. The January issue of the Scottish Alliance Bulletin is almost entirely devoted to the subject - the keynote of each article (and there is quite a number of them) is that printers should meet frequently and make a practice of discussing their difficulties more openly.

(Report of Secretary, Costing Committee, Minutes, January 1930).

*

The Chairman submitted the report of the Sub-Committee as follows: The Sub-Committee is of the opinion that there is no real cure for price-cutting except by means of further education along the lines of Costing, Estimating, and Production Records. They recommend that the suggestion of Mr. Bisset of a system of free check estimates be tried for a limited period of six months. They further recommend that the scheme submitted by the Costing Secretary whereby small firms could be assisted with their Costing figures be submitted to the Alliances for their consideration and report.

Mr. Crowlesmith urged that the recommendation on Price-cutting should be made more definite. In his opinion extensive and vigorous propaganda on simple lines should be undertaken and measures adopted for personal contact with members in every Alliance. Considerable discussion ensued, and finally the following resolution was agreed upon: That although some price-cutting can be stopped by closer co-operation between printers and by better salesmanship in the direction of creative and suggestive work, the best remedies will be found in educational propaganda along the lines of Costing, better estimating, and a study of production records. Further procedure was referred to the Sub-Committee.

(Costing Committee, Minutes, September 1930).

*

COSTING : WHY ?
(1931)

For nearly twenty years in and out of season, in the face of much tradition and greater prejudice, the organisation has kept well in the forefront of its activities the supreme necessity for proper costing methods. But there still survives, and every now and then gets thrown up to the surface, some ancient and primitive method of finding the cost, which proceeds by way of adding to wages sundry items of so-called expense, either by stages or in on jump. As capital charges vary according to operation, and expenses differ according to area required, wages alone are not a suitable basis for distribution of expenses. But tradition dies hard, and prejudice is more difficult to overturn than to establish principle where the mind is open.

The organisation has provided the services of experienced and trustworthy cost accountants for which it charges fees that are relatively nominal and merely a recognition of the assistance that can be given yet some say even these fees are too high!

The printer's product is peculiar and needs particular treatment. It cannot be measured by the barrel or the ton; it has neither length nor area, nor cubical contents; every unit differs from another. Thus it is impossible to find the cost of production by many of the standards adopted as gauges in other trades. These peculiarities render it absolutely essential that its cost be found by a system governed by principles that can be universally applied. Statistical records are a necessary prelude to cost recovery and the securing of a fair and equitable profit.

The Federation Costing System has been made as simple as possible within the bounds of necessity and accuracy - yet there are some who look askance at it and are unwilling to adopt it in its most simple form.

Why?

(Members' Circular, February 1931, p. 23).

*

To printers who refuse to install the costing system

❧ A good deal of misunderstanding still exists concerning the Federation Costing System, and this misunderstanding is not confined to buyers of Printing alone.

It Costs Money to Sell Printing

There is No Pact to Fix Prices

❧ It is true that many of our customers believe, or in many instances profess to believe, that those Printers who adopt the System enter into a pact to fix and maintain prices, and that the Federation has an elaborate organisation for disseminating information about the prices paid for certain printing orders. Let it here be stated without equivocation that those Printers

It Costs Money for Labour

It Costs Money to Deliver Goods

who adopt the Federation Costing System
do not enter into any pact (either written
or implied) to fix or maintain prices, and
that the Federation does not possess, and
has never possessed, any machinery for
collecting and distributing information
about the prices paid for printing orders.

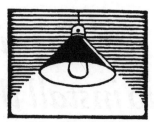

It Costs Money for Light

It Costs Money for Heat

There is No Obligation to Charge Fixed Hourly Rates

But what of the Printers who refuse to
consider the Costing System ? Some think
that if they adopt it they will be compelled
to charge specified rates for Composition,
Machining, Binding, etc., and they argue
that if they attempt to put these rates into
operation they will lose their customers,
and that the work will go to a Printer who
refuses to consider the Federation Costing
System and will not have it installed.

It Costs Money for Insurance

It Costs Money to Collect Money

The Federation Costing System Supplies the Printer with the Facts of His Own Factory

❧ Let us endeavour to make clear to the Customer and the Printer what exactly has been the work of the Federation Costing Committee. In a simple phrase it has been purely educational. The Committee has merely insisted on the importance of each member ascertaining his own costs. What the member does with these facts when he has obtained them is his own business. Should he find that he has been selling below his costs, he is still at liberty to do so if he wishes to remain a philanthropist. The Federation Costing System supplies the printer with facts; with the facts of his own factory. It does not follow that these particular facts will be the costs of the man across the street. The System says that these are *your* costs. The costs of the

It Costs Money for Bad Debts

It Costs Money for Paper & Handling

It Costs Money for Slack Periods

It Costs Money for Your Salary

man across the street may be higher or, on the other hand, they may be lower.

Know Your Costs — that's all

❧ If knowledge is power, surely it is advantageous to have knowledge. It is far better to know what your Composition is costing than to guess. You may guess and lose a job or you may guess and be landed with an unprofitable order. The result, ultimately, is exactly the same.

❧ Before your customer fixes the selling price of a new article he first ascertains his costs. If he didn't, you would soon lose a customer for ever. The Federation only asks its members to do what every wise and prudent manufacturer is doing. It says ascertain your true costs before fixing your selling price, and if you cannot find the facts for yourself the trained Cost Accountants of the Federation will find them for you.

A. J. Bonwick

It Costs Money for Plant

Depreciation Costs Money

It Costs Money to Buy Ideas

It Costs Money for Mistakes

(*Members' Circular*, March 1931, insert).

A PROFIT
ON EVERY JOB
MAY BE SECURED
If you adopt the
COSTING SYSTEM

FOR years printers guessed at their costs and were lucky if they covered them. What they needed was a system, simple, accurate, elastic, and prompt in results ; a system which could be applied to any business and would ensure a true determination of the cost of every job.

Then came the Federation Costing System. Its ultimate success was assured. Numbers of printers have adopted it, and over a period of years they have proved its value and usefulness. Not one of them would be without it. The system is available to every printer, and we want every printer to adopt it.

Those who now decide to receive the benefits of the Federation Costing System will have all the advantages of a highly developed system which has only been evolved after years of experience.

A PROFIT ON EVERY JOB?

*The Federation Costing System shows at once just where losses are occurring
and enables them to be avoided in the future*

Ours is a complex industry with many processes and a diversity of
materials. Accurate knowledge of costs is therefore absolutely essential. Few other trades are so fortunate in having a uniform system, and
are therefore unable to make comparisons of costs so successfully as
can the printing industry.

Seeing that the printer sells his product before it is manufactured, it
is essential that he should know his costs before fixing his selling price.
This involves the collection of details of cost as well as records of output. One without the other is ineffective, and neither can operate alone.

The principles of the Federation Costing System show how to arrive at a true cost by determining the Labour Cost per hour (or other unit) of each and every operation, whether time work or piece work, and by their application EVERY expense is allocated proportionately, thus making it possible to provide a means for recovery.

THE PRINCIPLES SHOW THAT

Cost Equals

(1) *Wages, plus Departmental and General Expenses.*

(2) *Materials, plus expenses of Buying, Storing and Handling.*

The Hourly (or Unit) Cost Rate must cover direct wages, a share of indirect wages, and overhead expenses.

The operations of the Federation Costing System show just where leakages are, and enable them to be rectified.

The system is a STANDARD one, adaptable and equally useful for small as for large businesses.

INSTALLATION AND MAINTENANCE

The Federation provides an efficient service for members in the direction of installation, tuition, and revision. Its trained Cost Accountants have had long and varied experience, and are in consequence familiar with most of the problems affecting cost.

SECRECY

Any figures disclosed to the Federation Officials are treated with absolute confidence and secrecy.

FIXING THE SELLING PRICE

THE "PRICE TO CHARGE" *consists of*

Wages, Materials and Expenses

[*To determine the proportion of which the Federation Costing System is essential*]

plus a Reasonable Profit

No business can be maintained without a reasonable profit. The printing industry is soon affected by any change in the volume of business—it is subject to abrupt rises and falls. Unfortunately, many short-sighted customers still regard printing as an expense rather than an investment, and in consequence competition is rife, with a resultant narrowing of the margin of profit. Fixing the selling price therefore needs the application of much common sense. The printer should be recompensed not only for his direct expenditure, but for his brains, capital invested, and technical knowledge.—*W. L. Bemrose.*

Recover Your Costs

(*Members' Circular*, April 1931, insert).

COSTING : TO-DAY'S NECESSITY
(1932)

THERE has arisen a real necessity for a keener perception of the value of accurate costing methods as an aid to management, and there are signs that not only is the need being felt, but action is being taken to remedy the fault.

Economic pressure has during the last few years steadily and persistently forced down the prices of every kind of product; no industry has been able to avoid the pressure. How far the producer has allowed circumstances to influence his actions, or whether he could have offered more resistance, need not be discussed here. The fact remains that the prices of many commodities have reached a level which must eventually force out of business many firms who are financially weak.

This pressure is one of the causes that must inevitably force the producer who wishes to escape disaster to examine every item of expenditure in order to ascertain whether the amount can be justified, and even whether the item is necessary.

Without a costing system the ultimate effect of even an attempt at expense reduction can only be surmised; the effort will be as the gropings of a blind man in strange surroundings, but without the aid of his highly sensitive nerves.

Costing is the science of measurement in terms of time and money. To attempt to run any kind of manufacturing business, particularly one as complicated as that of printing, without a costing system, is unwise, if not positively dangerous.

To say, as some do, that they cannot get work at their present costs, and that a costing system is therefore of no use, is begging the question. How do they know what their costs are? How can they know whether the expenses are being properly applied to the costs of production if the incidence of these expenses has not been considered?

What is the use of attempting to apply overhead expenses by means of a *flat* percentage on wages, as some do, when the actual variation in the different departments is anything from 80 per cent. to 350 per cent., according to capital value of the plant in use?

A non-costed business may be run successfully (that is from the point of view of profits) for many years, but a spell of bad trade will assuredly shake it to its foundations even if it does not destroy it. On the other hand, a business with a costing system kept thoroughly up to date is able to direct and control with confidence, because the important facts relating to departmental costs and earnings are known not only in total but in considerable detail also. It was a wise man who said "the little foxes stole the vines."

Whatever may be *thought* of the ability, or the lack of it, of some printers in regard to salesmanship, it can be *said* that printers as a body know their craft and can produce work of a high quality. This intense interest in the practical side of the industry has brought about changes in the methods of production that have led to a great increase in the amount of capital invested in plant.

These changes, whilst they may have had no immediate effect on the volume of the wages paid during normal periods, have had repercussions during these times of lessened demand for the printer's product. Manual labour is being displaced by the machine, at a higher cost per hour, although the cost per unit of production may be less. What needs to be constantly remembered is that work can now be produced more rapidly, and unless the flow of the work can be increased to meet the capacity of the machine the periods of standing time will be longer, and the losses greater.

Although it may be done reluctantly, it is not impossible to reduce wages to the level of the demands, but capital once expended must be constantly employed or loss is inevitable.

This gradual mechanisation and the increase in the speed of production has made scientific costing an absolute necessity, and it is the height of folly to invest capital and instal plant unless the apparatus for financial control is at the same time made equally efficient.

The printing industry was the pioneer in applying scientific and uniform costing methods, and is still the only industry that has any claims to continuity of effort, but it has still much to learn from the data such costing methods are capable of teaching.

Let those who have any consideration for the future of the industry think deeply about the circumstances of the present, and put behind them the temptations to grasp at what can only be a shadow, a snare, and a delusion; let them find their costs and stand by them.

(Members' Circular, January 1932, pp. 11-12).

*

MISUNDERSTANDINGS
(*1934*)

Although evidence is still available to prove that more member-firms are looking into their methods of finding the cost of production, there is still in some quarters a definite misunderstanding of the aims of this Committee, of its purpose in publishing hourly rates, and of what the Federation Costing System is intended to accomplish. This statement does not mean particularly that there is a new generation to be educated, but that the efforts of the last twenty years have not yet overcome that lack of desire to change over to a proper Cost-finding System from what is, at best a mere book-keeping method. The method of applying a percentage to so-called manufacturing cost, referred to by the Chairman at the Scarborough Congress, is far too prevalent.

The Costing Secretary has attended two meetings of considerable interest:-

1.- Lancashire & Cheshire Executive at Manchester.- This was a novel kind of meeting, held on the suggestion of Mr. Andrew. Its purpose was to afford every member an opportunity of (a) criticising the methods of the Federation Costing Committee and also the Costing System, (b) asking questions relative to the Federation Costing System and (c) making helpful suggestions. At the beginning references to price-cutting were ruled "out of order." There was little criticism, partly I believe, because the Alliance has an active Costing Committee; questions there were in plenty, and some helpful suggestions. The meeting was well attended and evoked much interest. A letter was sent by the Costing Secretary to each of those who attended the meeting suggesting that if more information were desired or assistance required he would welcome an inquiry.

2.- Group Meeting in the South Western Alliance at Newton Abbot. This meeting was well attended by members, some of whom had travelled considerable distances. The subject, chosen by the Alliance, was "The Federation Costing System: its simplicity, its guidance, and its value." The meeting showed its interest by close attention, and by the questions raised later. By calling on some of those present, in their own offices, later in the week, two installations and two revisions were secured, and there seem to be good reasons for expecting others.

(Report of the Secretary, Costing Committee, Minutes, March 1934).

*

TAKING STOCK OF THE COSTING SYSTEM
*(Address by W.H. Andrew, Chairman of the Costing Committee, at
the Thirty-Fourth Annual Congress of the BFMP, 1934)*

I think the time has arrived when it is desirable to take stock, as
it were, of the position of the Federation Costing System, and to
consider what is the attitude of the average printer towards that System.

Does the average printer really understand the aims and objects
of the Federation Costing Committee when it endeavours to persuade
him to adopt the System ?

Some of the thoughts in the minds of printers may seem strange
and unreal, almost untrue, when I mention them.

My only defence in putting them forward is that they are
statements made to members of our Costing staff when they pay their
visits in the course of their work.

You would hardly think it likely that in the minds of some
members there is a fear that in some way or other, if they put in the
Costing System, their businesses will pass out of their own control and
they will be dictated to by the Committee in London.

I suppose that fear is caused by an apprehension that some
official in London may see the firm's figures - those figures which they
have always endeavoured to keep within their own bosoms !

A more understandable notion is that of another group - that no
system can be applied to every kind of business, and that their own
business is all so peculiar that a peculiar system is the only one that
can serve their purpose . . . and some of the systems that are found in
use are peculiar !

Yet a third group are terrified at the thought that the figures and
calculations they are accustomed to will all be dispersed, as it were, in
the twinkling of an eye, and that they will be left without any prop to
support them, or any peg on which to hang their lute then denuded of
its strings.

A fourth group comprises those who have gathered here and
picked up there one idea and another, and formulated a method of
arriving at so-called cost which in ordinary circumstances as perhaps
served a purpose, but fails to act when matters become sub- or
abnormal. This kind of system is usually kept in the head of its creator,
and ceases to function when, from any cause whatsoever, he is absent
from business.

These groups usually have a method of finding cost which has a close relationship to the lines on which the Profit and Loss Account of a business is drawn up. There are sections representing the cost of wages and material, to which are added so-called establishment charges, followed by another amount representing gross profit. In few cases is there any attempt made - in most cases it would be impossible - to make a comparison between so-called cost and the actual cost until the balance sheet for the year is made available. The inherent weakness of this method is a fundamental one.

There is yet another group which lays claim to having invented a better system than the official one. Few, indeed, of these so-called "improvements" can withstand even a cursory examination, and any that can pass muster are usually not improvements but additions, creating more clerical work without commensurate increased utility.

TWO AIMS

Having defined the attitude of mind of typical groups of the Federation membership, let us consider for a moment or two what are the aims of the Federation Costing Committee. They are mainly two:-

(1) Every member to be a *user* of the official System
(2) Every user to apply the principles and methods uniformly.

To achieve the first of our aims is the more difficult of the two, and I have already outlined some of the causes which stand in the way of progress. There is a background of fear to these causes, with here and there probably a little misplaced pride. Fear of this kind is the offspring of ignorance, and can only be overcome by education, reinforced by the testimony of those who have adopted the system and use it properly.

On behalf of the Federation Costing Committee, I can state most emphatically that these fears of dire happenings are quite groundless. The Committee has no power, and desires none, to apply force of any kind. Its responsibility is to *persuade* members to adopt practices which are in the best interests of their businesses, because it firmly believes that the influence of the industry will be stronger as its members improve their business efficiency.

The second aim, that of the application of uniform methods, is less difficult of attainment but not less important. This is a problem which is being, or soon will have to be, faced by every industry in the country. It is being recognised with increasing force that comparisons between sections of all industries, and the individual businesses in these sections, must be made possible if efficiency is to be secured. Any attempt at comparison with data compiled on different bases is futile, and the industry with the best Costing System in universal use will prove to be the most efficient. We have had a long start - we were the pioneers of uniform costing methods - we have planned the roads, but we have had to overcome great obstacles, and our only weapon has been that of persuasion, and, as in all educational work, whilst our progress has not been startling it has certainly been steady and continuous.

TWO RESPONSIBILITIES
While the Federation Costing Committee is responsible for other work, these various activities are subsidiaries to the two great aims just recited.

The officials of the Costing Department have two great responsibilities:-

(a) To observe the strictest reticence in connection with those private details which they must of necessity handle if they are to render the best service. They neither report on these matters to the Costing Committee nor divulge what they may see and hear to other persons with whom they come into contact either in the course of their duties or in private intercourse.

(b) To see that the System is uniformly applied and thoroughly understood. When the System is applied and understood the user is absolutely free to use his judgment and business acumen, and is in a position to fix his own selling price with proved costs before him.

FOUR NAMES
This is the majority year of the Federation Costing System, and the necessity for change in the System during the course of these years has been slight. This in itself is the greatest possible testimony to the ability of those early Costing pioneers, of whom the greatest was the late W. Howard Hazell. There are still several survivors whom we had hoped to have with us to-day, in order to pay our tribute of honour to their work in the past, but increasing years and other difficulties have prevented them accepting our invitation.

We are, however, pleased to have with us on the platform this morning the youngest member of that historic group in the person of Mr. R. A. Austen-Leigh, whose ability and qualities the Federation has recognised on many occasions and in many ways.

Another member of the 1913 Committee who seems to look as young to-day as he did then is Mr. F. H. Bisset, and I am very glad indeed to announce that he also will address you.

Among so many that I could name I should like particularly to mention the late Mr. A. E. Goodwin. I have always thought he was singularly happy when he was doing Costing work. No matter how difficult a problem he might have to deal with as Secretary to the Federation, you had but to offer him a Costing problem and it was indeed with him "the world forgot."

He brought his human touch and a certain delight to Costing that only a real artist could do, and while he was not a member of the original Committee, I deem it a privilege to bring his name and his work back to memory.

(*Members' Circular*, July 1934, pp. 223-225).

*

THEN AND NOW OF THE FEDERATION COSTING SYSTEM
(address at the Thirty-Fourth Annual Congress of the BFMP by F.H. Bisset, 1934)

That was in 1913? What of the band of 1,200 people who assembled at Kingsway Hall? Where are the gallant 1,200 to-day? Where are their Costing Systems?

On the question of "Then and Now" I am, I think, fortunately or otherwise, in a peculiarly privileged position to judge, because from 1917 till I came to the Federation Secretaryship in 1929 I had no point of contact with the printing industry except an occasional casual meeting with A. E. Goodwin and others of my old friends in the industry - a hiatus of twelve years. And the thing which surprised and disappointed me most in coming back to the industry was the lack of substantial progress of the Costing System, measured in terms of installations. It is just as surprising, just as disappointing, to me to-day; indeed, more so, because the census which the Costing Department is at present undertaking disclose, so far as it has gone, a deplorable condition of things.

ACCUMULATED KNOWLEDGE

I personally quite fail to understand the position. An industry with a system so simple yet so scientific, so accurate, so flexible, so inexpensive to work, so efficient in every way - not merely as costing - and also, let me add, so interesting; and yet there are literally thousands of printers who are not making use of it. Indeed, there is even more to it than all that, because, while the efficiency and effectiveness of the Costing System in itself have been increasing as the original points of difference have been cleared up, we have had the Costing Secretary, Mr. Williamson, since 1917-for seventeen years, that is - accumulating daily and hourly experience, and Mr. Hull, for his somewhat shorter period of service, accumulating experience of printing businesses and their ways. We have a staff spending their whole lives - day in, day out examining printers' methods. They are bound, even if they were less intelligent men than they are, they are bound as a result of that experience to acquire an enormous mass of first-hand expert knowledge and experience of immense value to every printer who cares to make use of it. That does not mean in the least that they are going from one shop to another, giving away the personal secrets or experiences or methods of other printers. It does mean that they have acquired an all-round equipment which enables them to suggest sources of loss or leakage, improved methods, useful economics, in short, a tightening-up of the whole organisation of the particular printer on whose costing work they happen to be engaged. I see it actually happening in the daily correspondence as it passes through the office.

Then and now ! Yesterday we have been hearing about. To-day I have just been talking about. What about to-morrow, for that is the thing that matters? What are we going to do about it? Mr. Austen-Leigh has told us that in the earliest days three rival costing systems had to be dealt with, and that in the end there was a kind of amalgamation of the best points of the three.

WHAT IS WRONG ?

I will put three rival conundrums, and suggest replies - impertinent replies !

What is wrong with the Costing System ?

What is wrong with the Printer ?

What is wrong with the Costing Committee ?

What is wrong with the Costing System ? I should be inclined to say that there is nothing wrong with the Costing System, but that what *was* wrong with the System was that it passed through a certain period when it was inclined to be a little high-brow, mysterious, almost hush-hush. It has emerged from that.

What is wrong with the Printer? Just the old old story - the most fundamental of all human attributes, mental inertia. The slow progress of the Costing System is the greatest blot on the intelligence of the printing industry.

What is wrong with the Costing Committee? The Costing Committee has recently discovered what *was* wrong with it, and is taking steps to put it right. To-morrow we are devoting a whole session to the question of Salesmanship. Tomorrow we are laying on this table the latest contribution to improved salesmanship for printers in our new book, "Salesmanship for Printers" (it is an astonishingly fine book, by the way). And what has been wrong with the Costing Committee has been simply lack of salesmanship. The Costing Committee, to its credit, has discovered the weakness, and is setting out to change all that. And our main job this morning is to enlist your assistance towards deciding how the Costing System is to be sold, sold to the printing industry.

We need not have any conscientious scruples about selling the Costing System. At one of our Alliance week-end gatherings I met a printer who told me that he did not have the Federation Costing System, but he had its pup! After he had given me some details of his system, I had perforce to ask him who sold him the pup ! It is not a "pup" we have to sell to the industry, but something clean-bred, true-bred, sound in wind and limb and all the rest of it, something which the master printer simply cannot afford to do without.

(Members' Circular, July 1934, pp. 230-231).

*

CENSUS OF MEMBERS USING THE COSTING SYSTEM
(1934)

PROPAGANDA SUB-COMMITTEE REPORT *Census*. - A further report was given as to the census of members using the Costing

System. At the first issue of the census card 3080 cards were sent out and of these 838 were returned. Reminders to the number of 2246 were later sent out and 344 cards were returned. The total number returned was 1182 and of these 524 indicated use of the System, representing 17% of the total cards sent out. There were 21% of non-users, being 658 in number. Many of the non-users had asked for information to be sent them by post.

Reported further that the Propaganda Sub-Committee had that morning decided to revise a circular-letter which it was the practice to send to inquirers with specimen forms. A covering letter in the name of the Chairman thanking inquiring members for returning the cards and hoping they would instal the System later is proposed to be sent with the circular-letter and forms. It had also been decided to send to the Alliances the names of those members who had returned the cards, showing those who were users and those who were not; also to advise Alliances of those to whom the letter referred to would be sent.

A discussion ensued as to a suitable way for Alliances to deal with the matter and sundry suggestions were made. It was left to the Costing Secretary to frame a plan for communication with the Alliance Secretaries and to make suggestions on the lines of the discussion.

(Costing Committee, Minutes, September 1934).

*

PROGRESS IN THE COSTING MOVEMENT
(*1936*)

SUPPLEMENTARY REPORT FOR CHAIRMAN, VICE-CHAIRMAN, AND FEDERATION DIRECTOR

The following report has been prepared in order that some information should be available in the event of discussion arising as to the relative merits of quantity and quality for expediting progress in the Costing movement:

I recently extracted from our card index the names of a number of firms in whose businesses I personally had installed the System during the four years 1926-30 but who had never subsequently called upon our services. I am satisfied that I did everything possible during the time available to ensure that the System would be carried on and prove beneficial. With what result? One of the Cost Accountants, who has during the period covered by my report to the Costing Committee called upon each of those firms, reports that in: 88% of them the System has entirely lapsed, 7% are using the same figures which I gave them now nearly ten years ago, 5% claim to be making their own revisions but gave no proof of so doing.

One firm ordered the System to be re-installed, but a disappointing feature of this inquiry is that according to the Cost Accountant's reports only 14 per cent of those visited are now worth following up. This means that in these and all similar cases our task is harder than it was ten years ago for we have been in and failed. The reasons most frequently given for dropping the System were:

(1) We cannot get your prices.

(2) The information received is not worth the clerical work involved.

The Cost Accountant's own comments as a result of these visits, which I may say my own experience corroborates, can be summarised as follows:-

(a) A poor standard of book-keeping is all against us.

(b) The simplicity of the System has been over-emphasised, and demands for accuracy have had to be sacrificed to justify this.

(c) The doctrine of elasticity has not convinced the smaller and medium-sized business that a System which claims to cover the ramifications of a large office are anything but unwieldy for them.

(d) The delusion that the Federation Costing System is solely for the assessing the selling prices must be wiped out.

(e) The utmost keenness was apparent in all Y.M.Ps. [Young Master Printers] encountered, and any time that can be given to them is well worth while.

(f) Something really simple is the only solution for the very small office.

In connection with the last comment I am convinced of the possibilities of the suggestion which I put forward about three years ago and again some six months ago, but, whatever form it takes, that something is absolutely essential, particularly for use in conjunction with the postal installation scheme.

(Secretary, Costing Committee, Minutes, December 1936).

*

CANDID COMMENTS
A COSTING DISCUSSION
(*Federation Cost Congress, 1937*)

The Chairman: . . . And now, just a few words about the Costing
System . . . (interruption)

. . . AND DISCUSSION ENSUED

Mr. Past: It is useless, Mr. Chairman, for you Costing people to talk;
you will never do it.
Chairman: Do what, Mr. Past?
Mr. Past: Get every printer to charge the same price for his product. It
is a brilliant idea on the part of you big people to get us to raise our
prices so that you pinch our work, but we're not standing for it;
besides, we should be accused of forming a ring.
Mr. Shaw: Mr. Chairman, we are not attempting to set every printer to
charge the same price. Our objective is that the cost of every job done
shall be found by the most accurate method. Experts are agreed that the
directly chargeable hour is the best unit of cost, and also that part of
the expenses must be recovered on materials and outwork - labour
should not be expected to carry all the burden.

Customers nowadays consider that accurate costing is the reverse
of forming a ring - they welcome it, and they are willing to place
contracts on the cost shown by the use of the Federation Costing
System, plus an agreed percentage of profit.

(Members' Circular, July 1937, p. 254).

*

COSTING-DUAL FUNCTIONS
INSTALLATIONS AND REVISIONS - EDUCATION
AND SKILLED ADVICE
(*W.H. Andrew, Chairman of Costing Committee, at the Federation*
Cost Congress, 1937)

We have no front-page news to offer, and yet I think this year I
can sense a feeling of a change that is coming over the policy of the
Costing Department. Perhaps I had better describe it not as a change
of policy but as a greater realisation of the dual requirements we have

to meet.

Our ideal, of course, is that every printing office should have the Costing System installed, followed by periodical revisions. When, however, the number of new installations is reported to the quarterly Council meetings, the criticisms may be and has been that at the present rate of progress it would take us a few hundred years to achieve our ideal. And we have to admit that the criticism is sound. In practice the installing of, say, 4,000 systems or even a proportion of those is beyond the possibility of our staff, just as much as it is obviously unlikely that we should get those 4,000 orders in even a hundred years.

HEALING THE SICK

Let us consider for one moment a medical school of a great university-say, London. Suppose the staff at that school had to do all the operations even in London and attend to all the medical cases. It would be an impossibility, of course. But what that medical school does is to train students who go out into the world to perform those operations and attend to the sick.

Now whilst the Costing staff does undertake all the installation operations they are able to do, we are realising to-day more than ever the importance of our educational work of training up our young doctors-in fact, our young costers.

Our costing examinations-what are they but an aid to training cost accountants ? Our linking up as far as possible with all the schools of printing in the country. Our co-operation with Messrs. Pitman in the establishment and running of their Correspondence course. Our reissuing a comprehensive book on Estimating - to the authors of which I paid tribute at the Annual Meeting. Our re-drafting and re-writing the text book on Costing. The vast importance of the issue of hourly rates.

(Members' Circular, July 1937, p. 303).

*

SEARCHLIGHT ON COSTING
(1938)

A definite trend towards intensive education of costing staffs has become increasingly noticeable in the past few years. It has manifested itself in collaboration with technical schools, the introduction of the Federation Examination in 1933, the first Teachers' Conference in

1935, culminating with the series of week-end courses by the Federation Chief Cost Accountant. The courses began in the natural home of the good Experiment, the Midlands, and are being introduced to the other Alliances in turn.

In the jargon of the Costing Department they are "intensive" courses. No more suitable term could be found, as the experience of some twenty-five years is compressed into fourteen hours of explanation, dissertation, discussion, and even argument between two o'clock on Saturday and eight o'clock on Sunday. Interjection with consequent mild sidetracking prevents staleness as a sprayer prevents set-off, and it is a remarkable fact that students are as keen at the close as they are at the commencement.

By the end of May some 160 students will have crowned a winter's study with fourteen hours of conference, led by an acknowledged expert. It is difficult to gauge the usefulness of this arrangement; the misunderstandings eliminated, the problems solved, and the keenness engendered by the frank personality of A.D. Hull.

The type attracted to these week-ends is mainly the cost clerk, with a leavening of principals and Y.M.Ps [Young Master Printers]; in short, the type who will give up a week-end in the mid-morning of the year to study what could be a prosy subject. The very success of these week-ends and the fine type attracted to them will ensure their continuation and they should become a Spring movement as regular and as well anticipated as the Annual Congress.

(Members' Circular, June 1938, pp. 170-171).

*

SUGGESTED GUIDE TO ALLIANCES ON THE MEMBERSHIP AND FUNCTIONS OF COSTING COMMITTEES
(1938)

MEMBERSHIP

The Committee should consist of one representative from each Association in the Alliance and such members should have an active interest in the Federation Costing System. It is suggested that members should be selected from those whose firms have the system installed and working in their own offices, or from others qualified to serve by previous experience, actual knowledge, or special training. The Alliance representative to the Federation Costing Committee should be on this committee and in many cases he will be the suitable member to act as Chairman of it.

FUNCTIONS

1. To deal with all costing questions within the Alliance and to report and make recommendations to the Alliance Executive and through it and its own representative, to the Federation Costing Committee.

2. To consider, prior to the Executive Meeting, the report of the representative to the Federation Costing Committee and to suggest to the Executive the best way of putting into effect any proposals made.

3. After the meeting of the Executive, if it so desires, to carry out in detail such proposals within the Alliance.

4. To take a live interest in existing Costing Classes within its area and to encourage the establishment of Costing Classes in all Technical Schools within the Alliance, to make known other sources of instruction, including the Correspondence Course and to encourage entry to the Federation Costing Committee's Examinations.

5. To conduct propaganda through the Alliance Journals and in any other way considered likely to be effective and to seek new ideas and angles of approach whereby the Costing System may be advanced.

6. To stimulate the interest of Associations in Costing and, in conjunction with the Federation Costing Department, to arrange for instruction, demonstration and interview within their areas and to encourage associations to form their own local Costing Committees.

7. To offer to the Federation Costing Committee such criticism of its work as may arise within the Alliance.

8. It is an advantage if an Alliance Costing Committee in co-operation with the Association can compile lists of those working the system (a) fully and (b) in part and especially if they can make out a list of printers likely to receive favourably a visit from a Federation Cost Accountant asking to put in the Costing System.

It is emphasised that the proper function of Alliance and association Costing Committees is to deal with Costing problems and the subject of selling prices should be excluded from their discussions as far as possible.

It is also suggested that members should from time to time be encouraged to open discussion on some subject connected with Costing at Alliance Costing Committees, Alliance Executive Meetings and at Association Meetings.

The Alliance and Association Costing Committees should give to their Secretaries all possible help and encouragement in their work of furthering the Costing movement in their area.

(Costing Committee, Minutes, November 1938).

*

COSTS AND WAR
Costing Service Scheme for War-time
(*1940*)

A new and improved scheme of costing service to members has been devised by the Costing Department. This service will be of special value to members in meeting the ever-changing conditions of war-time.

War conditions make it more than ever necessary for the printer to know his costs. When the prices of materials are altering daily, especially when the movement is upward, and when labour conditions are undergoing frequent adjustment owing to the calling-up of men for military service and other reasons, the need for accurate costing is increased.

The printer who studies costs to-day is materially assisted in meeting changing conditions, and in planning ahead and shaping policy. Further, he is helping to conserve his business and maintain it on a sound basis for future development.

The new service scheme offered by the Costing Department is designed to give the utmost help to every member of the Federation. It maintains and simplifies contact between the member and the Costing Department, and ensures continuity of service, advice, and consultation. Full details are given below. Members are invited to study the scheme, and to decide now to avail themselves of the new service from its inception.

THREE SERVICES

To meet the varying requirements of Printers the scheme provides a choice of three services:

SERVICE " A " - Complete (items I to XII).

SERVICE " B " - Partial (items I to VIII only).

SERVICE " C " - Advisory (items I to IV only).

SERVICE "A" (Complete)

I. A periodical bulletin on matters of topical interest, the effect of current events on cost, questions of common interest which have been dealt with by the Costing Department since the last bulletin was issued, etc. It will also contain current average figures such as, for example, the ratio of (a) various expenses to total cost, (b) indirectly chargeable hours to directly chargeable hours for each Department, (c) production to wages paid in each Department, etc. Such figures have been found to be invaluable for purposes of comparison.

II. The calculation of an hourly cost rate for any machine or operation not listed on the Standard Hourly Rate Card.

III. All the assistance with any costing or estimating problems which it is possible to give by correspondence.

IV. The examination of an estimate or cost sheet, and provision of a certificate if required for presentation to the customer in cases of dispute.

V. A quarterly examination of and comments upon all costing records, provided these records are sent to the Costing Department compiled and totalled in accordance with Chapter IV of the tenth edition of the Textbook.

VI. A visit by a Cost Accountant once a year for consultation and advice on costing problems, to help the cost clerk with any difficulties or to render other assistance in keeping the costing working smoothly and with full efficiency.

VII. The preparation of a Statement of Net Surplus or Deficit (Textbook, Chapter V, paragraph 128) - an assessment of net profit or loss - at any time, provided the costing records are compiled and totalled in accordance with Chapter IV of the Textbook and sent to the Costing Department on a form supplied for that purpose.

VIII. The reconciliation of an assessed net profit with audited accounts at the end of a financial year.

IX. The preparation of a new Budget - Statement of Expenses - or adjustments to an existing one, as and when necessary.

X. The allocation of budgeted expenses to departments.

XI. The calculation or adjustment of hourly cost rates as and when necessary.

XII. Any reasonable service not covered by items I to XI.

SERVICE "B" (Partial)

This service comprises items I-VIII of the full service. There are some firms who, while they use the Federation Costing System and prepare their own budget and statement of expenses, and compile the various records, feel that they need help and expert advice in the different uses to which these records can be put. This need is met by Service B.

SERVICE "C" (Advisory)

This service comprises items I to IV of the full service. It covers the needs of those firms who thoroughly understand all the details of the Federation Costing System or use an equivalent system of their own, but who would like to keep in touch with the Costing Department, because of its publications, statistics, and the estimating service, so that they can benefit from the department's recommendations and experience, and obtain figures for comparison with their own.

(Members' Circular, April 1940, pp. 73-75).

*

COSTING COMMITTEE POLICY
(Report to the Federation Council, 1946)

An extensive memorandum setting out in detail proposals and suggestions for the future of the Committee and the Costing Department was circulated to the Alliances in the latter half of 1945. Alliance Costing Committees have given careful consideration to that memorandum and have submitted comments and criticisms which have been examined by the Costing Committee.

The Committee now submits to the Council for approval a brief statement of the lines on which it proposes to work during the next few years.

I General

(a) The Committee will continue to impress upon the industry the need for sound costing and informed estimating.

(b) It is of the opinion that a long-term policy of education in the value of costing is the best means to that end.

II Education in Costing

The Committee has reviewed the various methods adopted in the past to spread a knowledge of scientific costing throughout the industry. It proposes to revise the Costing Text Book from time to time as need arises, to publish a simplified text book on costing and book-keeping, and to encourage the formation of students' classes on costing wherever possible. Liaison between the teachers of costing and the Department's staff will be maintained and extended, and the Annual Examinations will continue to be organised by the Department. Certificates and Diplomas will be awarded to successful students. The Costing Secretary and his staff will assist in the running of short courses in costing subjects at the request of any of the Alliances.

III Services of Costing Department Staff

The Committee is of the opinion that the Costing Secretary should have more time free in future to devote to the educational side of the Committee's work and to matters of broad principle affecting the whole of the membership. The staff is now being built-up to pre-war level in order to meet an increasing demand for individual service on revision and installation of the Federation Costing System.

A revised statement on Costing Service available to members, free and on a fee-basis, is appended to the Report.

IV Information Service

The Committee proposes to encourage the Department to increase and extend the scope of the valuable statistical records that have already been compiled. Standard Hourly Cost Rates will be revised and re-issued to all members from time to time, and reports on general increases or decreases in cost will be circulated as and when necessary. Information and advise on members' individual problems will be freely given on request, and the staff of the Department will continue to advise representatives of the Federation in negotiations and discussions with H.M.S.O. and local and public authorities.

V Estimating

The Committee has in hand the revision of the Federation Text Book on Estimating. It is proposed to encourage the formation of students' classes in estimating, and liaison between the Department and teachers of estimating will be developed. Arrangements are in hand to increase the capacity of the Department to provide an estimating service to members. Free check estimates are available to members and independent estimates at an appropriate fee will be prepared. The Committee intends to stimulate an interest in the compilation of production records by members with a view to the analysis of such records and the publication of production figures in due course.

VI Propaganda

The Committee looks forward to the full co-operation of Alliance Costing Committees in its efforts to promote the wider use of scientific methods of costing throughout the industry. Assistance in providing speakers for meetings will be given as far as possible, and notes and charts will be made available where necessary. Shortage of staff in the Department is impeding speedy progress at present, and a large increase in demand for individual service would be embarrassing, but the shortage will be remedied in a few months time, and the Committee expects to be able to meet all reasonable demands from the membership on the subjects of Costing and Estimating early in 1947.

(Costing Committee, Minutes, April 1946).

*

FIFTY YEARS OF COSTING
(1963)

(a) *Inquiry regarding use of Federation Costing System.* Replies to the questionnaire have so far been received from 1,072 firms. . . Returns indicate that the Federation Costing System is being fully used by 53 per cent of the firms which have replied. In these firms a periodic statement of net profit or loss is produced from the cost records.

Although the figures reveal a satisfactory use of the Federation Costing System in well over 500 businesses there is an evident need for continued intensive efforts to encourage many other members to use sound costing and cost control methods.

This year will provide an excellent opportunity for intensified efforts in that visits by cost accountants to members firms to discuss costs and costing methods, and to give on-the-spot advice, are being arranged in several alliances.

Arrangements are also being made to meet the requests of 240 firms which in their replies to the questionnaire stated that they would

like a Federation cost accountant to call on them.
(Costing Committee, Minutes, March 1963).

*

FIFTY YEARS OF COSTING: The Chairman stated that despite the loss of staff during the course of the year the Costing Department had been engaged in a great many additional activities in connection with the Jubilee Year of Costing, including the following:

Special calls by staff on member firms | 150
Talks given by members of staff | 30
Articles written | 10
Profit Planning and Control conferences | 6
Special meetings arranged | 16
Hourly Cost Rate calculator on the B.F.M.P.
      stand at IPEX
Business Session at Dublin Congress
  - PROFIT FROM PRINTING.

He thought it could be said that the year had been a success and that much had been done to create in the minds of members an awareness of the need for satisfactory profits and adequate financial cost controls to achieve good results. These efforts should not be regarded as being completed, however, for there was still an immense amount of work to be done. It had been evident throughout the year that the re-styled activities of the department, which paid closer attention to profit and profitability than in the past, had been considerably more effective. The work started this year would continue, and would be intensified wherever possible.
(Costing Committee, Minutes, December 1963).

*

POSTSCRIPT

In the wake of the Jubilee Year of the Federation Costing System, the Costing Committee of the BFMP continued its efforts to advance printers' awareness of costs, prices and profits. In 1963 the issue of price competition and the need to convince printers of the 'economic fallacy of cutting' still featured on the Costing Committee's agenda. This persistent echo from earlier times contributed to the longevity of the Costing Committee's campaign. By 1964 it was reported that: the Costing Department was staffed by ten cost accountants; costing examinations for printers were to take place in Britain, Africa and India; 20,000 copies of Federation books on costing and estimating had been sold in the last decade; and that enrolments on correspondence courses in costing and estimating since 1957 had exceeded 800.

During the 1960s the work of the Costing Department of the BFMP was extended beyond purely costing issues. A management ratios scheme, management advisory services for members and a computer service were all introduced. In 1967 this broader emphasis was reflected in the reconstitution of the Costing Committee as the Management Accounting Committee and the Costing Department as the Management Accounting Section. The remit of the Committee was now to advance the efficiency and profitability of the printing industry by dealing with "all matters relating to accountancy, costing, estimating, management ratios, and related services to management" (Management Accounting Committee, Minutes, November 1967). The functions of the Management Accounting Committee were defined as follows:

1. To encourage the use of modern management systems and techniques at all levels of management and to provide, where possible, the means by which the necessary knowledge and skills can be acquired.
2. To offer guidance to members by making recommendations in matters of principle and in methods of approach to particular problems.
3. To assist members to become more profitable by providing information and advisory services which meet current needs.

4.   On behalf of members to negotiate with Government
          Departments and other bodies (Management Accounting
          Committee, Minutes, September 1971).

In November 1970 it was reported that in the last fifteen months
598 member firms located in England and Wales had availed
themselves of at least one of the services provided by the Management
Accounting Section (management accounting, management ratios,
computer service, salary survey).

Following a review on the constitution and structure of the
BFMP in 1971, the minutes of the Management Accounting Committee
cease to be available. However, the British Printing Industries
Federation (BPIF) has continued to publish literature on costing and
management. The thirteenth edition of *Estimating for Printers* (which
was first published in 1916) appeared in 1989. The twelfth and last
edition of the costing manual which first appeared in 1913 was
published in 1956. This was replaced by *Cost Accounting for Printers
Part I* in 1971 and *Costing for Printers Part II* in 1973. In 1988
*Getting the Measure of Your Business. How to Use the BPIF
Management Information System* appeared. This contains an
instructional handbook, a workbook and specimen forms in a style
reminiscent of the early costing manual. The promotional literature for
the latter also contains a statement which is redolent of an earlier age:

> No business can hope to be successful unless management
> has a firm grip on costs. Control depends on information,
> and a proper system is particularly essential in the printing
> business.

BIBLIOGRAPHY

The Accountant.

Alford, B.W.E. (1965), "Business Enterprise and the Growth of the Commercial Letter-Press Printing Industry, 1850-1914," *Business History*, Vol. 7-8, pp. 1-14.

Banyard, C.W. (1985), *The Institute of Cost and Management Accountants: A History*, ICMA, London.

The British Printer.

The Caxton Magazine.

Child, J. (1967), *Industrial Relations in the British Printing Industry. The Quest for Security*, George Allen & Unwin, London.

Clegg, H.A. (1976), *The System of Industrial Relations in Great Britain*, Basil Blackwell, Oxford.

Clegg, H.A., Fox, A. and Thompson, A.F. (1964), *A History of British Trade Unions Since 1889*, Clarendon Press, Oxford.

Costing Committee, Minutes, 1913-1967, British Federation of Master Printers (St Bride Printing Library, London).

Farnham, D. and Pimlott, J. (1990), *Understanding Industrial Relations*, Cassell, London.

Fleischman, R.K. (1996), "A History of Management Accounting Through the 1960s," in Lee, T.A., Bishop, A., and Parker, R.H., *Accounting History From the Renaissance to the Present*, Garland Publishing, New York, pp. 119-42.

Fleischman, R.K., Mills, P.A. and Tyson, T.N. (1996), "A Theoretical Primer for Evaluating and Conducting Historical Research in Accounting," *Accounting History*, NS Vol. 1(1), pp. 55-75.

Gospel, H. and Littler, C. (eds.) (1983), *Managerial Strategies and Industrial Relations*, Heinemann, London.

Howe, E. (1950), *The British Federation of Master Printers, 1900-1950*, Cambridge University Press, Cambridge.

ICWA (1944), *Uniform Cost Accounting and the Principles of Cost Ascertainment Schemes*, Institute of Cost and Works Accountants, London.

The Lithographer.

Loft, A. (1986), "Towards a Critical Understanding of Accounting: the Case of Cost Accounting in the UK, 1914-1925," *Accounting, Organizations and Society*, Vol. 11, pp. 137-70.

Loft, A. (1988), *Understanding Accounting in Its Social and Historical Context. The Case of Cost Accounting in Britain, 1914-1925*, Garland, New York, NY.

London Members' Circular, Master Printers and Allied Trades Association, *Members' Circular* (London).

Management Accounting Committee, Minutes, 1967-1972, British Federation of Master Printers (St Bride Printing Library, London).

Members' Circular, The Federation of Master Printers and Allied Trade Association.

Mitchell, F. and Walker, S.P. (1995), "Another Costing Revolution," *Management Accounting*, December, pp. 36-37.

Mitchell, F. and Walker, S.P. (1997), "Market Pressures and the Development of Costing Practice: The Emergence of Uniform Costing in the UK Printing Industry," *Management Accounting Research*, Vol. 8(1), pp. 75-101.

Moore, M. (1970), *How Much Price Competition ?*, McGill University Press, Montreal and London.

Most, K.S. (1961), *Uniform Cost Accounting*, Gee, London.

Most, K.S. (1977), *Accounting Theory*, Grid Inc., Columbus, OH.

Musson, A.E. (1954), *The Typographical Association. Origins and History up to 1949*, Oxford University Press, London.

Napier, C.J. and Carnegie, G.D. (1996), "Editorial," *Accounting, Auditing & Accountability Journal*, Vol. 9(3), pp. 4-6.

O'Brien, D.P. and Swann, D. (1968) *Information Agreements, Competition and Efficiency*, MacMillan, London.

Palmer, G. (1983), *British Industrial Relations*, Allen & Unwin, London.

Powell, L.M. (1926), "Typothetae Experiments with Price Maintenance and Cost Work," *Journal of Political Economy*, Vol.34, pp. 78-99.

Report of the First British Cost Congress (1913), Federation of Master Printers of the United Kingdom of Great Britain and Ireland, London.

Scherer, F.M. (1970), *Industrial Pricing: Theory and Evidence*, Rand McNally College Publishing Company, Chicago.

Sessions, M. (1950), *The Federation of Master Printers, How it Began*, William Sessions Ltd., London.

Solomons, D. (1950a), "Uniform Cost Accounting - A Survey, Part I," *Economica*, August, pp. 237-53.

Solomons, D. (1950b), "Uniform Cost Accounting - A Survey, Part II," *Economica*, November, pp. 386-400.

Walker, S.P. and Mitchell, F. (1996), "Propaganda, Attitude Change and Uniform Costing in the British Printing Industry, 1913-1939," *Accounting, Auditing & Accountability Journal*, Vol. 9(3), pp. 98-126.

Wells, M.C. (1978), *Accounting for Common Costs*, Center for International Education and Research in Accounting, Champaign, IL.

T - #0187 - 101024 - C0 - 229/152/13 [15] - CB - 9780815330240 - Gloss Lamination